D0982585

Rhythmic Phenomena in Plants

Second Edition

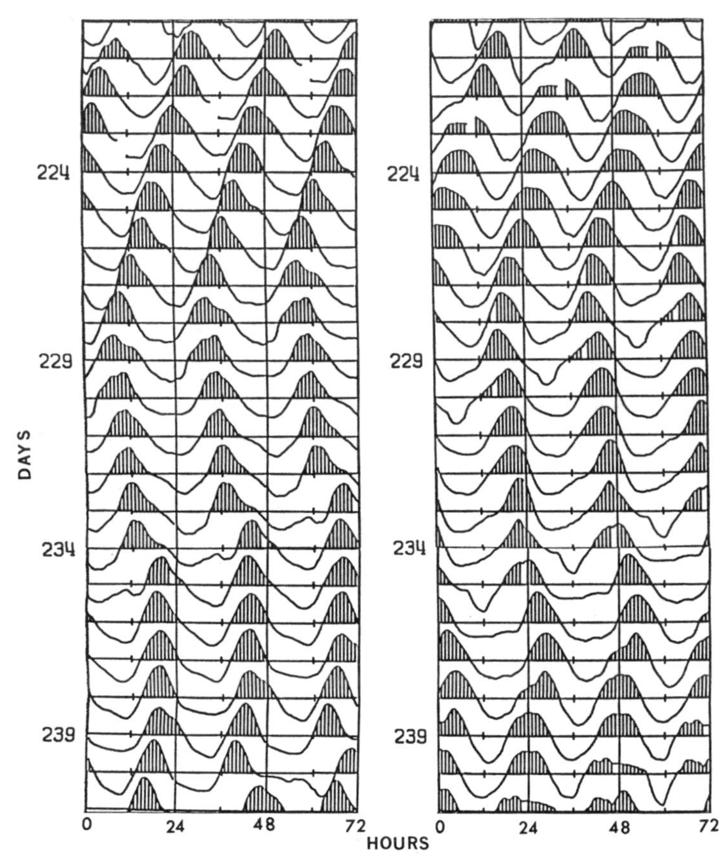

Recordings of the rhythms in chloroplast migration (left) and electrical potential difference between tip and base (right) in the same young *Acetabularia chalmosia* cell without a cap in constant light at 20°C. Days 221 to 241 are shown in triple plot including the preceding, the current, and the following days on each line. Note that the period in constant light sometimes changes and that darkening the cell from 10 to 18 h on day 235 caused a phase shift in the next day. Shaded areas: above average concentrations of chloroplasts (left) and electrical potential difference (right). (Courtesy of H.-G. Schweiger.)

Rhythmic Phenomena in Plants

Second Edition

BEATRICE M. SWEENEY

Department of Biological Sciences
University of California
Santa Barbara, California

ACADEMIC PRESS, INC.
Harcourt Brace Jovanovich, Publishers
San Diego New York Berkeley Boston
London Sydney Tokyo Toronto

ACADEMIC PRESS, INC.
1250 Sixth Avenue, San Diego, California 92101

United Kingdom Edition published by
ACADEMIC PRESS INC. (LONDON) LTD.
24–28 Oval Road, London NW1 7DX

Library of Congress Cataloging in Publication Data

Sweeney, Beatrice M.
Rhythmic phenomena in plants.

Includes bibliographical references and index.
1. Biological rhythms in plants. I. Title.
QK761.S95 1987 581.1'882 87-1075
ISBN 0–12–679052–3 (alk. paper)

PRINTED IN THE UNITED STATES OF AMERICA

87 88 89 90 9 8 7 6 5 4 3 2 1

To my husband, Paul H. Lee, who introduced me to the wonders of WordStar, dBase II, and SuperDex and in every way encouraged the rewriting of this book.

Contents

7 Rhythms That Do Not Match Environmental Periodicities

8 Biological Clocks and Human Affairs

Preface

When thinking about and writing this book, I have imagined before me a young student who has heard that plants and animals can tell time and has come to ask how such a clock works. I cannot really answer this question. No one knows the answer yet. What I have done is to set forth the evidence that indeed biological clocks exist and, in fact, are very common in eukaryotes. This book is about the evidence from studies of all kinds of plants, from unicellular algae to flowering trees.

There is a good reason to begin a study of biological clocks by considering plants rather than animals, although there are many examples of rhythmic phenomena in animals including ourselves. Plants and animals in general follow the same rules with respect to rhythms, although animals appear to have a higher order clock in the brain. Animal physiology is largely the study of the interaction of organs rather than cells. Many of the rhythms known in animals are rhythms in activity, which are not easy to understand at the cellular and molecular level. In plants, with their much simpler organ systems, it is easier to arrive at the level of cells and molecules. It is there that we must look for the mechanism of the biological clock.

Perhaps best of all as experimental material are the unicellular organisms in many of which clocks can be seen clearly. Single cells that can live free, making their components using light energy trapped in photosynthesis, may be the best of all for the study of biological timing. For this reason, I have described these systems in particular detail, hoping to entice new biologists interested in studying the biological clocks to use these organisms.

In a book of this size it is not possible to include all the interesting experiments on rhythmic phenomena, even all those on plants. There are now hundreds of papers on this subject. I have been forced to pick and choose, influenced of course by my own research interests. To compensate in part for this limitation, I have included long lists of references with each chapter, so that students will know where to look further.

The chapters have come out to be of very different lengths: the long ones are those describing the experiments on circadian rhythmicity. This reflects the fact that we know more about these rhythms than about rhythms with other frequencies and reflects my own personal interest as well. I justify this choice by the thought that eventually the study of all rhythmic phenomena will profit from a thorough understanding of one.

In writing and illustrating this book I have had help from a number of investigators whom I gratefully acknowledge here. In particular, I would like to thank Drs. Erwin Bünning, Victor G. Bruce, Leland N. Edmunds, Jr., Wolfgang Englemann, Jerry F. Feldman, Arthur W. Galston, Karl C. Hamner, J. Woodland Hastings, Noburô Kamiya, Dieter G. Müller, John D. Palmer, E. Kendall Pye, Frank B. Salisbury, Ruth L. Satter, Hans-Georg Schweiger, Kendrick C. Smith, Vera Vielhaben, Rütger A. Wever, Malcolm B. Wilkins, and Rose Zimmer, all of whom have contributed illustrations. I would also like to acknowledge the American Association for the Advancement of Science, the American Society of Plant Physiology, Il Ponte Publishers, Plenum Press, and Brookhaven National Laboratories for permission to reproduce figures from their publications.

Beatrice M. Sweeney

Rhythmic Phenomena in Plants

Second Edition

1

First Observations: The Patterns of Plant Movement

To us self-important humans, it is rather startling to realize that animals can tell time without the aid of wristwatches. The thought that even plants can accomplish this feat is at first almost too much to believe. However the evidence is quite clear. Plants as well as animals show behavior patterns that repeat with a definite and accurate time interval without any cues from the outside world. They must then have time-keeping devices by which they regulate their activities with respect to the important time intervals in their environment—the day, the tide, the month, and the year. Such innate timing ability was first demonstrated in plant sleep movements.

The power to move is so firmly associated with animals that we feel a sense of surprise the first time we notice that plants too can move. Yet we are familiar with the fact that the cells of plants and animals are fundamentally alike. How much more astonished must the naturalists of long ago have been when they observed that leaves and flowers could change their position. Perhaps the earliest record of plant movement comes from Pliny's writings (see Brouwer, 1926) recording the observation that some kinds of leaves take up a different position at night from that which they occupy during the day. Carl Linnaeus, in 1755, coined the name "plant sleep" for these movements (Fig. 1.1), and they are called sleep movements to this day.

That sleep movements represent anything other than a direct response to light during the day and darkness at night was first suspected by the astronomer De Mairan in 1729. To test his idea, he placed sensitive plants (probably *Mimosa pudica*) in darkness for several days and saw that, indeed, the movements continued and hence the change in leaf position did not require a change in the light conditions. This finding was confirmed by Zinn in 1759 and, in 1825, by De Candolle, who studied the leaf movements of *M. pudica* and found that, if the plants were placed in artificial light every night, the leaves still "slept" and "woke" as they would have done had the plants been growing under the natural alternation of day and night. There was, however, one difference. Now the time from one maximum spreading of the leaflets until the next was not 24 h as it would have been in nature, but shorter than this by 1–1.5 h. No particular importance was attached to this observation of a shorter period until much later. De Candolle also looked at plants kept continuously in darkness, where he found the movements present but somewhat less regular than in continuous light. He observed that the leaves of many other plants also showed sleep movements, *Oxalis* and *Phaseolus*, for example. Even some flowers that close at night continued this behavior in the absence of changes in the light environment. In his studies of the sleep movements of *Mimosa*, De Candolle also found that it is possible to reverse the course of the movements by placing the plants in darkness during the day and in artifical light at night. Thus, it seemed that, although the changes in light were not the cause of the leaf movements, light and darkness still influenced them.

There now began a period of intense interest in the patterns of motion of leaves, tendrils, and shoots that touched almost every eminent plant biologist of the second half of the nineteenth century: Darwin (1876), Dutrochet (1837), Hofmeister (1867),

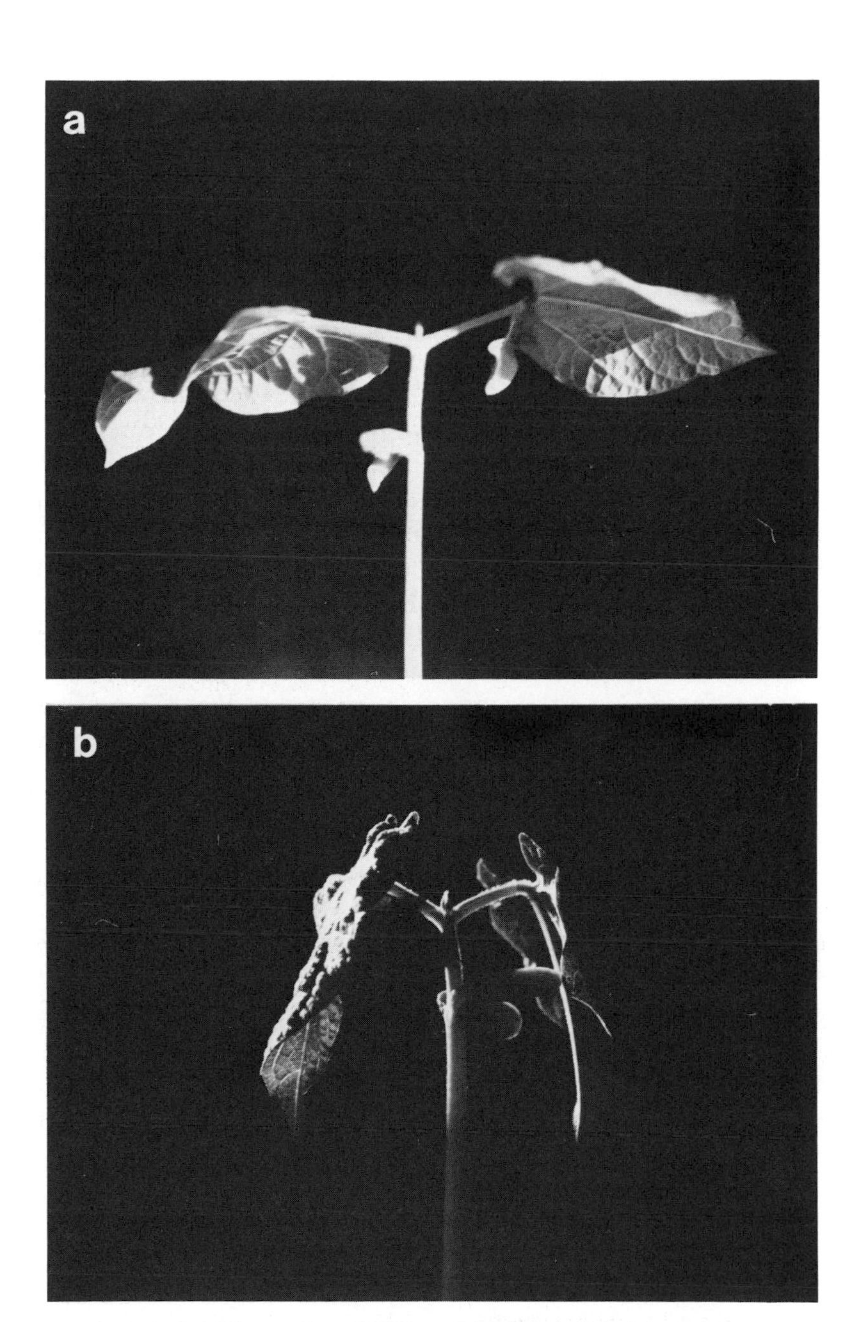

Fig. 1.1. Seedlings of the garden bean *Phaseolus* photographed during the day (a) and during the night (b) to show the difference in the position of the leaves at these two times, because of the "leaf sleep movements."

Sachs (1857), Pfeffer (1875). These men were concerned with how movements were executed, but also with why they existed and what caused their periodicity, whether movement sprang directly from some environmental periodicity or whether it had some other basis.

At first there were no methods for making automatic recordings of the movement of leaves, so that all observations depended on the wakefulness as well as the patience and tenacity of the investigator. Measurements were necessary, however, before anything more could be learned about sleep movements and their causes.

Darwin was the first to devise a system for recording plant movements (Darwin and Darwin, 1881). He attached a very thin glass fiber to the leaf or stem, on the free end of which was a wax bead. He marked the position of this bead on a glass plate in front of the plant, writing the time beside it. An example of one of his experiments using a tobacco leaf is shown in Fig. 1.2. Darwin was fundamentally a very careful observer rather than an experimentalist. He confirmed the widespread occurrence of sleep movements, rather than adding new data toward interpreting their cause.

Julius Sachs, on the other hand, was first and foremost an experimental plant physiologist. In April 1863 he selected a plant of *Acacia lophantha*, which he assures us had "new fine, very healthy-looking leaves," and placed it inside a wooden box with a thermometer beside it. He then measured the angle the leaflets made with the petiole every hour for 4 days, recording the temperature with each measurement. The leaves continued to show sleep movements for 2 days in darkness, but they were not correlated with the variations in temperature inside the box. Thus, the possibility was eliminated that, in darkness, temperature cycles were causing the sleep movements.

Pfeffer (1875) also studied the sleep movements of *A. lophantha*. He demonstrated that cycles persisted in constant light as well as in constant darkness (Fig. 1.3), although his lighting scheme (Fig. 1.4) left something to be desired with respect to constant intensity and temperature. Pfeffer showed that expos-

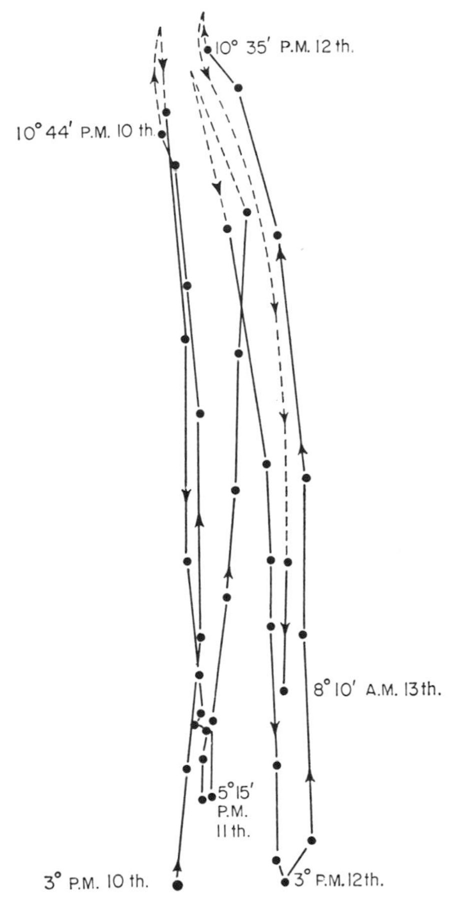

Fig. 1.2. Movement of a leaf in *Nicotiana tabacum* traced by Charles Darwin on a vertically-placed glass (Darwin and Darwin, 1881).

ing a plant that had lost its rhythmic sleep movements in long-continued constant light to a new light-dark cycle caused the sleep movements to resume. This new rhythm was brought about by placing the plant in darkness from 8 A.M. until 4 P.M. and was reversed with respect to natural day and night. Pfeffer (1911, 1915) considered the persistence of sleep movements to be "aftereffects" of the previous light–dark cycles. However, he

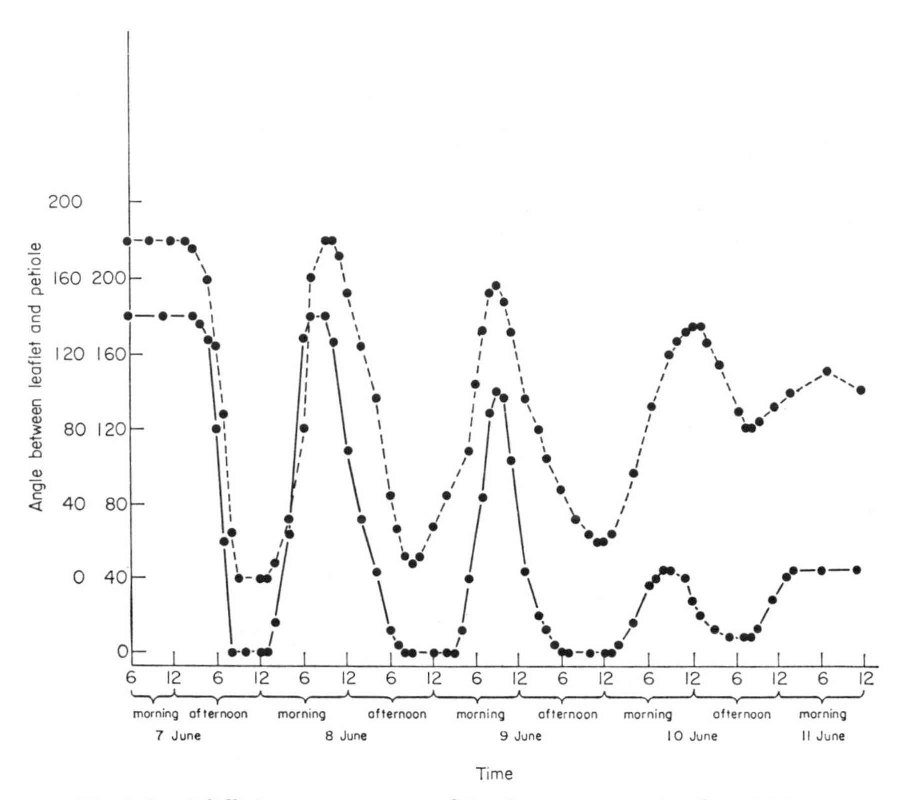

Fig. 1.3. Pfeffer's measurements of the sleep movements of an old (upper curve) and a young leaf (lower curve) of *Acacia lophantha* in continuous darkness (Pfeffer, 1875).

never performed the crucial test of this theory, which was to grow a plant under light–dark periods, the sum of which was not 24 h, then to move the plant to constant conditions and observe whether or not aftereffects reflected the imposed light–dark cycle. This experiment was soon performed by Semon (1905), who grew seedlings of *A. lophantha* in 6-h-long days and 6-h-long nights from germination and measured their sleep movements both in the 6-h light–dark cycle and after transfer to continuous light (Fig. 1.5). No aftereffects of the short days and nights were seen. Instead the sleep movements in continuous light showed a clear rhythmicity with a period of 24 h.

Pfeffer devised an ingenious method for recording leaf movements automatically. Figure 1.6 shows the arrangement of his apparatus, which is still sometimes used. Note that such an arrangement writes the record upside down, the pen dropping when the leaf rises. His new experiments convinced Pfeffer that Semon was right and that the leaf movements were indeed intrinsic, probably inherited.

Several other periodic responses in plants were discovered in the latter part of the nineteenth century. Sachs (1887) demonstrated the existence of rhythms in growth rate and in the exudation of fluid from the cut ends of shoots of plants. Periodicity in growth continued in constant darkness. Shorter and longer cycles than 24 h were also observed. Darwin (1876) noted that some plants execute jerky movements at short intervals of 4–6 min, while Pfeffer (1875) alluded to the rapid pulsation of contractile vacuoles in *Ulothrix* (12–15 sec) and *Gonium* (26–60 sec). He also discussed the annual changes in the rate of bud growth in trees and noted especially that pear trees transplanted from Europe to Java were known to become evergreen, but

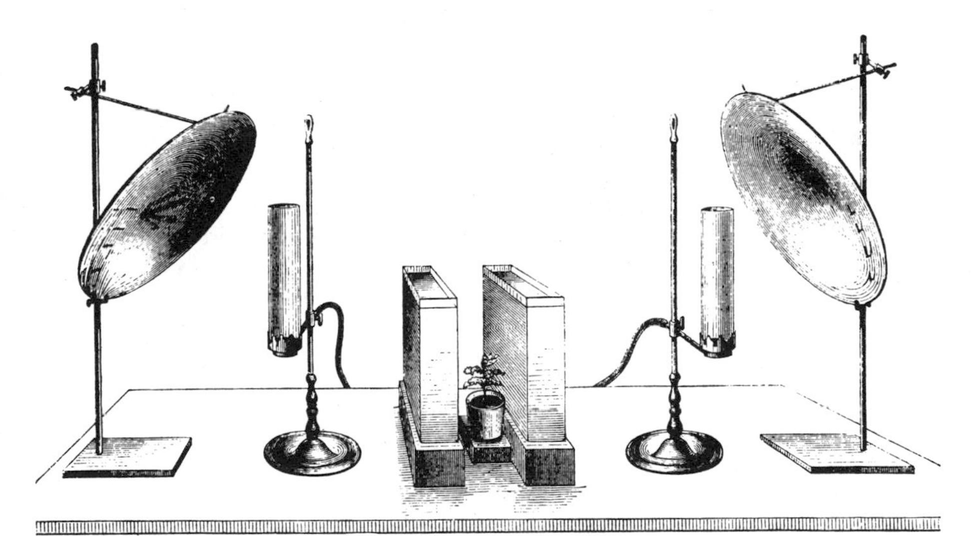

Fig. 1.4. Pfeffer's arrangement for lighting plants continuously (Pfeffer, 1875).

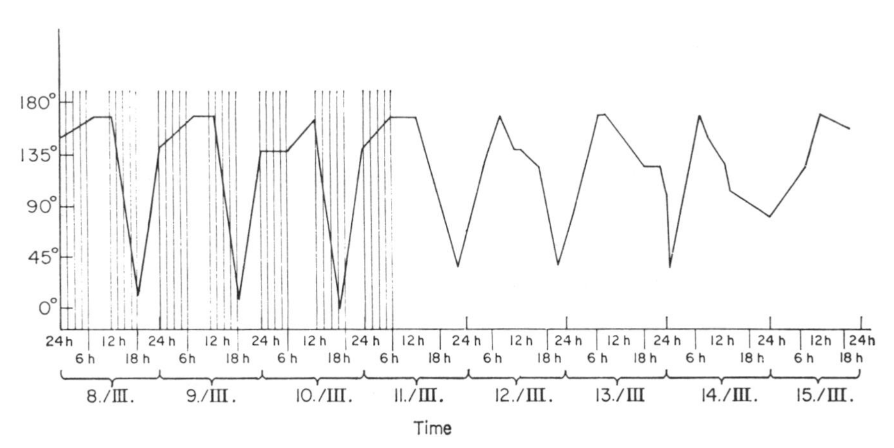

Fig. 1.5. Semon's measurements of the sleep movements of leaves of *Acacia lophantha* in LD 6 : 6 followed by continuous light, showing that after-effects of the light–dark regime are not detectable (Semon, 1905).

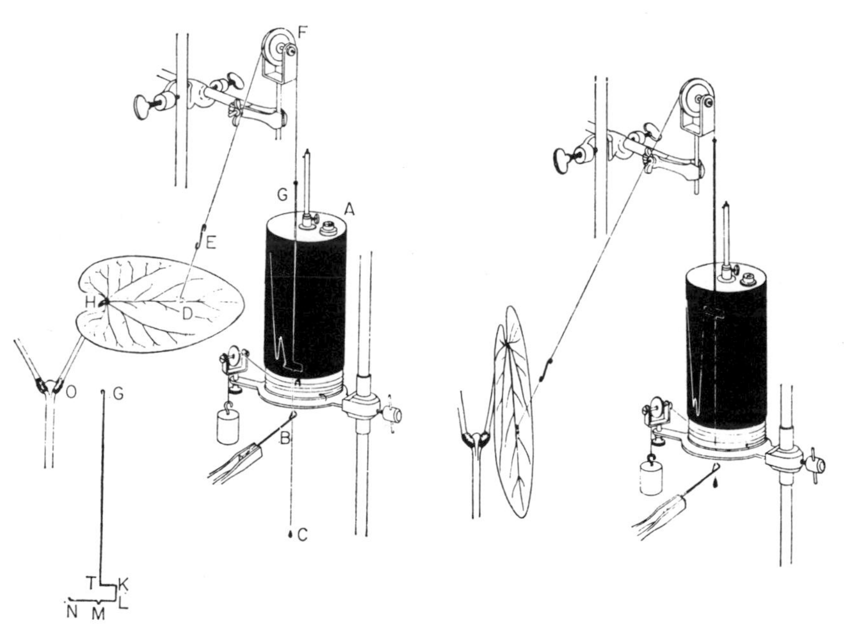

Fig. 1.6. Device for recording leaf movements automatically, similar to that devised by Pfeffer and used by Brouwer and Kleinhoonte. On the left the leaf is shown in the day position, on the right in the night position (Brouwer, 1926).

individual buds still showed rest periods not synchronized even with other buds of the same tree. Pfeffer also mentioned that native tropical trees lose their leaves and go through a short rest period that does not necessarily occur at the same time in adjacent trees. Considering these observations, Pfeffer speculated that there might be an hereditary annual periodicity independent of changing season.

The first period of intense activity in the study of plant periodicities came to a close with the admission by Pfeffer (1915) that leaf movements appeared to be inherited. Two papers, both of which contained beautiful automatic recordings of sleep movements of the leaves of the large bean *Canavalia ensiformis* appeared in Holland somewhat later, one by Brouwer (1926) and one by Kleinhoonte (1929). However, the interpretation of the data followed the same tradition.

Let us summarize the accomplishments of the plant biologists of the nineteenth century with respect to rhythms. Periodic leaf movements, and to some extent periodic growth and exudation, were measured under natural conditions and both in continuous light and darkness, with some attempt at constant temperature. The results of these experiments established that these cycles did not depend on environmental cycles but were generated endogenously and were inherited rather than learned. That light did have an effect on periodicity was known from experiments where day and night were reversed artificially, but the details of light effects had not been examined. No one even hinted that rhythms might be indications that plants can distinguish one time from another, and possess a chronometer useful in a number of ways.

Following the last papers of this period, there occurred a lull in research on rhythms that lasted without interruption for almost 20 years. Paradoxically, this quiescent period coincided with a leap forward in the technology necessary for control of the environment. Truly constant light and temperature conditions could now be achieved in the laboratory. Constant light was the order of the day. Physiologists strove to grow extremely uniform plants without variation from plant to plant or *from time to time*, and so they regarded variation in the time domain as a result of failure in their experimental technique. Such an attitude does not welcome the discovery of rhythmicity. Indeed, rhythmicity was not usually present since the favorite environment was one of constant bright light, a condition unfavorable for rhythms.

The papers of Bünning were harbingers of a renewed interest in periodicity, the first of which was published in 1930 with Stern. In the following letter Bünning tells in a very engaging way the story of how he became interested in the sleep movements of plants:

POTATO CELLARS, TRAINS, AND DREAMS: DISCOVERING THE BIOLOGICAL CLOCK*

Erwin Bünning

The story went something like this. At the Institute for the Physical Basis of Medicine in Frankfurt, the biophysicist Professor Dessauer (an X-ray specialist) became interested in the effects of the ionic content of the air upon humans. Those were the years when people began to be interested in atmospheric electricity, cosmic rays, and so on. Naturally, humans could not be used as experimental objects, and so in 1928, Dessauer searched for botanists to work on plants. One whom he found was Kurt Stern, who lived in Frankfurt; the other was me, who had just finished my doctoral work in Berlin. So we began in August of 1928 to contemplate the problem. In the process we came upon the work of Rose Stoppel, who had been studying the diurnally periodic movements of *Phaseolus* (common bean) leaves. In the process she had found, as had several other authors, that under "constant" conditions in the darkroom, most leaves reached the maximum extent of their sinking (their maximum night position) at the same time; namely, between 3:00 and 4:00 A.M. Her conclusion: Some unknown factor synchronized the movement. Could this be atmospheric ions? We had Rose Stoppel visit us from Hamburg for two or three weeks so that we could become familiar with her techniques. In our group we always called her *die Stoppelrose* ("stubble rose"), and the name was most appropriate. She was energetic and persistent, so persistent that she just died this January in her 96th year. Her results also appeared in our experiments: The night position usually occurred at the indicated time. Then we investigated the effects of air that had been enriched with ions or air from which all the ions had been removed. The result: Nothing changed—atmospheric ions are not "factor x."

*Letter dated June 2, 1970, published in the textbook by Salisbury and Ross (1978, p. 312); reprinted by permission.

We then decided that our research facilities at the Institute were insufficient. Hence, after *die Stoppelrose* had left, Stern and I moved to his potato cellar, where with the help of a thermostat, we obtained rather constant temperature. Contrary to the practice of Stoppel, who turned on a red "safe" light to water her plants, we went into the cellar just once a day with a very weak flashlight and felt around with our fingers for the pots and recording apparatus so that we could water the plants and so that we could see if everything was in order with the recorders. The flashlight was weakened with a dark red filter so that one could see only for a few centimeters' distance. In those days it was the dogma of all botanical textbooks that red light had absolutely no influence upon plant movements or upon photomorphogenesis. We did one other thing differently from Stoppel. Since Kurt Stern's house was a long way from our laboratory, we didn't make our daily control visit in the morning, rather only in the afternoon. The result: Most of the maximal night positions no longer appeared between 3:00 and 4:00 A.M., but rather between 10:00 and 12:00 A.M. Hence we concluded: The dogma is false. Red light must synchronize the movements so that a night position always appears about 16 hr after the light's action. That was "factor x." When we eliminated this hardly visible red light, we found that the leaf movement period was no longer exactly 24 hr but 25.4 hr. [This circadian feature of the clock was the key to understanding its endogenous nature—Ed.]

So that was about how that story went. Naturally, I could also tell you the story about how I came upon the significance of the endogenous rhythms for photoperiodism. That was about in 1934. Of course, I had already often asked myself how such an endogenous rhythm might ever have any selection value (in evolution), and I had already expressed the opinion in 1932 in a publication (Jahrbuch der Wissenschaftlichen Botanik 77: 283-320) that some interaction between the internal plant rhythms and the external environmental rhythms must be of significance for plant development. But there were coincidences in the story. As a young scientist, one naturally had to allow himself to be seen by the power-wielding people of his field, so that he might receive invitations for promotion. Hence, I traveled in 1934 from Jena to Könisberg in Berlin to introduce myself to the great Professor Kurt Noack. We discussed this and that. He mentioned that discoveries were being made that were so remarkable one simply couldn't believe in them. Such a one, for example, was photoperiodism. He, as a specialist in the field of photosynthesis, must certainly know that it makes no difference whatsoever what program is followed in giving the plant the necessary quantities of light. Then, as I was riding back on the train, the idea came to me—aha, for the plant it does make a difference at which time light is applied, if not exactly in photosynthesis, nevertheless for its development!

I could present a third story, one from very recent years. I have long felt

> that the daily leaf movement rhythms had no selection value in themselves. As you know, I have recently changed my mind. The movements could indeed be important in avoiding a disturbance of photoperiodic time measurement by moonlight. Before I had begun to test the idea experimentally or even to think about it, it came very simply one night in a dream. The dream (apparently as a memory of one of my visits to the tropics): tropical midnight, the full moon high above on the zenith, in front of me a field of soybeans, the leaves, however, not sunken in the night position. My thoughts (in the dream): How shall these plants know that this is not a long day? They had better hide themselves from the moon if they want to flower.

His work with the leaf movements of *Phaseolus vulgaris* revived interest in diurnal leaf movements, but far more important was his realization that these leaf movements might be manifestations of a fundamental ability to tell time. In 1920, Garner and Allard discovered that the flowering of tobacco was regulated by the length of the day, the phenomenon of photoperiodism. In order to respond to day length, it is of course necessary to measure this time interval. Bünning invoked the ability to tell time manifest in sleep movements to explain how the day length is measured in photoperiodism.

Like daily rhythmicity, photoperiodism was first discovered in plants. It soon became clear that animals also detected the seasons by measuring the length of day and hence could tell time. Now for the first time, daily rhythms in animals became of interest. Pittendrigh's studies of the eclosion in *Drosophila pseudoobscura*, the first paper in a long series appearing in 1954, defined the nature of this rhythm and its response to short pulses of light and emphasized that this rhythm was the overt manifestation of a biological clock that might be consulted for different purposes in the fruit fly and in other insects. The behavior with respect to temperature pulses was investigated and a theory postulating the existence of a driver and a slave oscillator was put forward (Bruce and Pittendrigh, 1957). The behavior of the timing of eclosion following short pulses of light was condensed as a phase-response curve (PRC), a presentation that made possible the comparison of the effects of light pulses on a variety of organisms. PRCs for a number of other circadian rhythms (Fig. 1.7) were found to be essentially similar to that for eclosion in *Drosophila.* The response to light pulses became one of the characteristic properties of circadian rhythms. Pittendrigh's analysis of his experiments with *Drosophila* initiated the modern work on circadian rhythms, well under way by 1960, when all the research to date was summarized at the Symposium on Quantitative Biology at Cold Spring Harbor Laboratory (Chovnick,

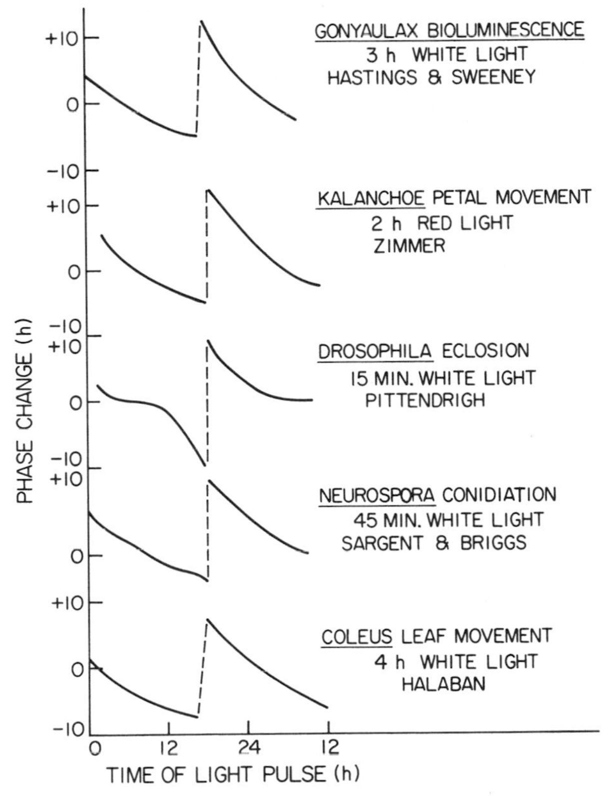

Fig. 1.7. Phase-response curves (PRCs) for five circadian rhythms in five different organisms. Notice how similar they are. The rhythm, the organism, and the treatment are written beside each curve. Advances in phase are positive, delays, negative. Time points are plotted in the middle of the light exposure (Sweeney, 1977). See pp. 22–23.

1960), an interesting volume to read to understand the state of the study of biological clocks at that time.

REFERENCES

Brouwer, G. (1926). "De periodieke Bewegingen van de primaire Bladeren bij *Canavalia ensiformis*." J. H. Paris, Amsterdam.

Bruce, V. G., and Pittendrigh, C. S. (1957). Endogenous rhythms in insects and microorganisms. *Am. Nat.* **91**, 179–195.

Bünning, E., and Stern, K. (1930). Uber die tagesperiodischen Bewegungen der Primarblatter von *Phaseolus multiflorus*. II. Die Bewegungen bei Thermokanstanz. *Ber. Dtsch. Bot. Ges.* **48**, 227–252.

Chovnick, A. (ed.) (1960). Biological Clocks. *Cold Spring Harbor Symp. Quant. Biol.* **25**, 1–524.

Darwin, C. (1876). "The Movement and Habits of Climbing Plants." Appleton, New York.

Darwin, C., and Darwin, F. (1881). "The Power of Movement in Plants." Appleton, New York.

De Candolle, A. P. (1825). "Physiologic végétale." Paris.

De Mairan, M. (1729). "Observation botanique, " p. 35. Histoire de l'Académie Royale de Sciences, Paris.

Dutrochet, H. (1837). "Mémoires pour servir à l'histoire des végétaux et des animaux." Brussels.

Garner, W. W., and Allard, H. A. (1920). Effect of length of day on plant growth. *J. Agric. Res.* **18**, 553–606.

Hofmeister, W. (1867). "Die Lehre von der Pflanzenzelle." Engelmann, Leipzig.

Kleinhoonte, A. (1929). Uber die durch das Licht regulierten autonomen Bewegungen der Carnavalia-Blatter. *Arch. Neerl. Sci. Exactes Nat. Ser. 3B* **5**, 1–110.

Linnaeus, C. (1755). Somnus plantarum. Amoenitat. *Academicae* **4**, 333.

Pfeffer, W. (1875). "Die periodischen Bewegungen der Blattorgane." Engelmann, Leipzig.

Pfeffer, W. (1911). Der Einfluss von mechanischer Hemmung und von Belastung auf die Schlafbewegungen. *Abh. Saechs. Akad. Wiss. Leipzig, Math.-Phys. K1.* **32**, 161–295.

Pfeffer, W. (1915). Beitrage zur Kenntnis der Entstehung der Schlafbewegungen. *Abh. Saechs. Akad. Wiss. Leipzig, Math.-Phys. K1.* **34**, 1–154.

Pittendrigh, C. S. (1954). On temperature independence in the clock-system controlling emergence time in *Drosophila. Proc. Natl. Acad. Sci. U. S. A.* **40**, 1018–1029.

Sachs, J. (1857). Uber das Bewegunsorgan und die periodischen Bewegungen der Blatter von Phaseolus und Oxalis. *Bot. Z.* **15**, 809–815.

Sachs, J. (1863). Die vorubergehenden Starre-Zustande periodischbeweglicher und reizbaren Pflanzenorgane. II. Die vorubergehende Dunkelstarre. *Flora (Jena)* **30**, 465–472.

Sachs, J. (1887). "Lectures on the Physiology of Plants" (H. M. Ward, transl.). Oxford Univ. Press (Clarendon), London and New York.

Salisbury, F. B., and Ross, C. W. (1978). "Plant Physiology," 2nd ed., p. 312. Wadsworth, Belmont, California.

Semon, R. (1905). "Über die Erblichkeit der Tagesperiode. *Biol. Centralbl.* **25**, 241–252.

Sweeney, B. M. (1977). Chronobiology (circadian rhythms). *In* "The Science of Photobiology" (K. C. Smith, ed.), Fig. 8-3, p. 216. Plenum, New York.

Zinn, J. G. (1759). Von dem Schlafe der Pflanzen. *Ham. Mag.* **22**, 49–50.

2

A Short Dictionary for Students of Rhythms: Parameters of Rhythms and How to Calculate Them

In any field there are certain ideas that recur very frequently. Sometimes the expression of these ideas in everyday language requires many words and awkward explanatory phrases. For this reason, a special vocabulary soon develops, a convenient shorthand that may be obscure to the uninitiated. In the field of rhythms this is particularly true, since the recurrent ideas are somewhat more than ordinarily difficult to express in a few words. Many single observations are required to show the presence of a rhythm and therefore one wishes to combine all these data in the form of a characteristic graph with time as the abscissa. Then, too, it is necessary to refer to the light conditions during the experiment, where a light period of say 12 h alternates with a dark period of the same or a different length and is followed by continuous light. Shorthand ways of expressing the conditions and the results of an experiment have been developed. A brief explanation of these terms seems worthwhile to include here.

Perhaps the first thing to define, so that we are quite clear what we are talking about, is the term *biological rhythm.* A *rhythm* is a regular fluctuation in something, e.g., in the position of a leaf like the rhythms described in Chapter 1, in the rate of a physiological process, in shape, in color, or in activity. The important thing is that a *repeating* pattern can be discerned clearly. However, not every repeating pattern is a rhythm. A rhythm is self-sustaining, and it is important to understand what that implies. Perhaps an example will make this clear. Let us say that we observe that the leaves of a tree in the garden are folded together every night but spread wide again every morning. There is a repeating and predictible pattern in this behavior, but is it a biological rhythm? We cannot tell, since the enormous difference between conditions during the day as compared with those at night may be *causing* the differences that we see in the leaves. They may simply open in light and close in darkness whenever these conditions occur. If this is the case, the change in leaf position is not a biological rhythm but a response to environmental light. The pattern in leaf position is not self-sustaining and would disappear at once were the environmental changes of day and night removed. Just this experiment must be done to detect the presence of a true rhythm: the oscillation must be shown to persist, at least for a time, in constant conditions of light and temperature.

Before we consider the terms that are used to describe rhythmic phenomena, let us pause to think about a repeating pattern with which we are familiar, the turning of a wheel. The name *cycle* comes from this analogy where a circle is changing its position in time, yet any point on the rim is always returning to contact the ground. There are several ways in which we can picture what is happening to points on this wheel as it turns. From a vantage point to the side of the vehicle, a point on the rim will appear to follow a scalloped course (Fig. 2.1a), while looking at the wheel from in back, the same point will describe a sine curve (Fig. 2.1b). Both these patterns repeat over and over as the wheel continues turning; they represent *oscillations.* Not all

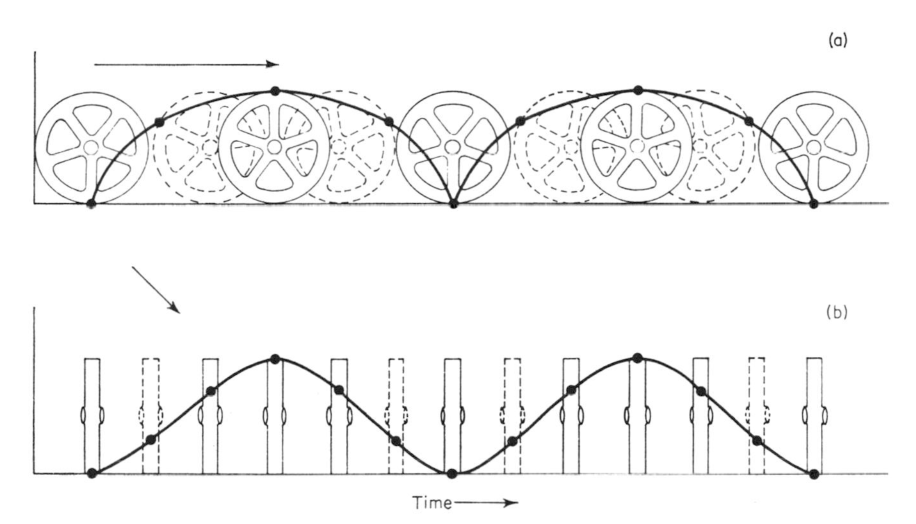

Fig. 2.1. Derivation of cycles from the turning of a wheel.

oscillations have the shape described by a sine curve, however, as we shall see later.

Rhythms can be plotted in a number of different ways (Fig. 2.2). Method *a* shows the most detail. Method *b* is usually used when a rhythm has only one distinguishing phase point per cycle, as do the activity rhythms of many animals where the beginning of activity is sharp but the middle and end of activity are rather indefinite. Method *c* is common in describing rhythms important in medicine when the time relationship between different rhythms is their most important feature.

The fundamentally oscillatory character of biological rhythms makes it appropriate to use for them the terms that describe physical oscillations, like that of a pendulum or an electron. The time required for an oscillation to make one complete cycle and return again to the same starting position is the *period.* It is the time for one revolution of the wheel or one circuit of a leaf from horizontal position back to horizontal again. Note that the period is the same, no matter what point in the cycle is used to measure it.

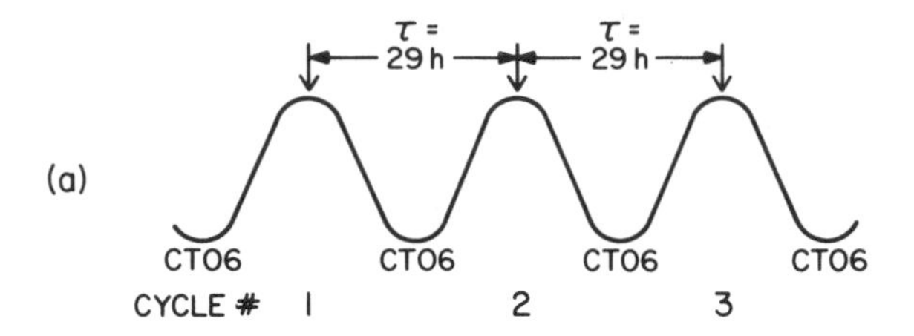

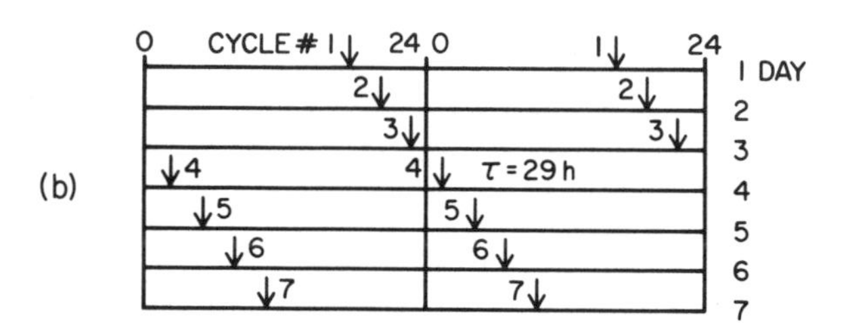

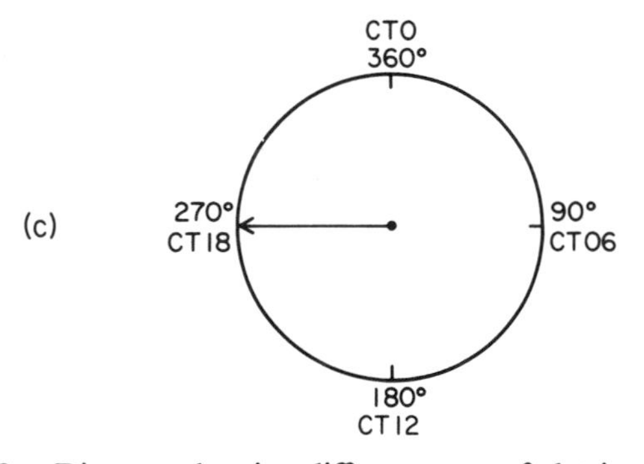

Fig. 2.2. Diagram showing different ways of plotting circadian rhythmic data. (a) Amplitude of the rhythm as a function of time. (b) Phase points plotted on a separate line representing each day, one above the next. The data are double-plotted for clarity. The slope of such plots gives the period. (c) Rhythmic data plotted as a cycle of 360°. The arrow denotes the time of the maximum.

A SHORT DICTIONARY DEFINING TERMS USED IN BIOLOGICAL TIMING*

Biological cycle A sequence of events in an organism that repeat in the same order and at the same time interval.

Endogenous, self-sustaining or free-running rhythm A cycle that persists under constant environmental conditions for at least several periods, usually much longer, before becoming indistinguishable or "damped out." If evidence for persistence is lacking, the cycle is known as a periodicity, for example "diel periodicity" observed in a light–dark environment where the period is 24 h.

Period (tau, τ) The time required to complete one cycle. Biological rhythms are classified according to their free-running periods, as follows:

ultradian—a period substantially less than about 24 h.

circadian—a period of about 24 h (hence "circadian" = about a day).

infradian—a period appreciably longer than 24 h.

semilunar—a period of about 14 days, or half a lunar cycle.

lunar—a period of about 28 days, the time between one full moon and the next.

circannual—a period of about (circa) a year.

annual—a period of a year.

Frequency (f) The number of cycles in a unit time or 1/period.

Amplitude (A) The degree to which the observed response varies from the mean, also sometimes used as the difference between peak and trough of a rhythm.

Phase (ϕ) An arbitrarily chosen part of a cycle, usually related to a part of another cycle, for example the *night phase* of a circadian rhythm, the part of the rhythm usually coinciding with darkness in the environmental cycle of day and night. One may also speak of the *day phase.* These terms apply equally well to cycles in constant conditions.

*For further discussion of terminology, consult Aschoff *et al.* (1965) and Palmer (1976).

phase angle difference—the angular difference between the phase of one cycle and that of another, when one cycle is considered to be 360°.

phase shift—a change in the phase of one cycle relative to that of control cycle. If the phase is advanced in time relative to the reference cycle, the phase-shift is positive (+); if the phase is delayed, the phase-shift is negative (−) (Fig. 2.3). (phase shift = reset.)

phase-response curve (PRC)—a graphic representation of a number of phase shifts caused by treatments of short duration relative to one cycle and positioned at different parts of the cycle (a pulse experiment). The abscissa of a PRC is the time of the pulse (beginning or middle), while the ordinate is the amount of the phase shift, either positive or negative (Fig. 2.3).

phase jump—the sudden change from delay phase shifts to advances in phase in a PRC.

type 1 and type 0 PRCs—the PRCs to very effective resetting signals such as bright light are called type 0 response curves since they vary around a zero slope or horizontal baseline; response curves to weakly resetting stimuli are called type 1, because their baseline has a slope of 1.

singularity—is a unique time and dose in a phase-response experiment, the result of which is arrythmicity. The existence of singularities implies that the oscillation is a limit cycle.

Circadian time or clock time (CT) One cycle is considered to be 24 h long, even if under free-running conditions the period is not 24 h solar time; hence 1 h is $\frac{1}{24}$ of the period. The choice of when a cycle starts at 00 h (= 24 h) is arbitrary. In the United States following Pittendrigh, 00 h is dawn, the end of the dark part of the environmental light–dark cycle or the phase corresponding to this point in constant conditions. In Europe, the convention of Bünning that 00 h is midnight or the corresponding phase is sometimes followed.

Subjective day and subjective night The phase correspond-

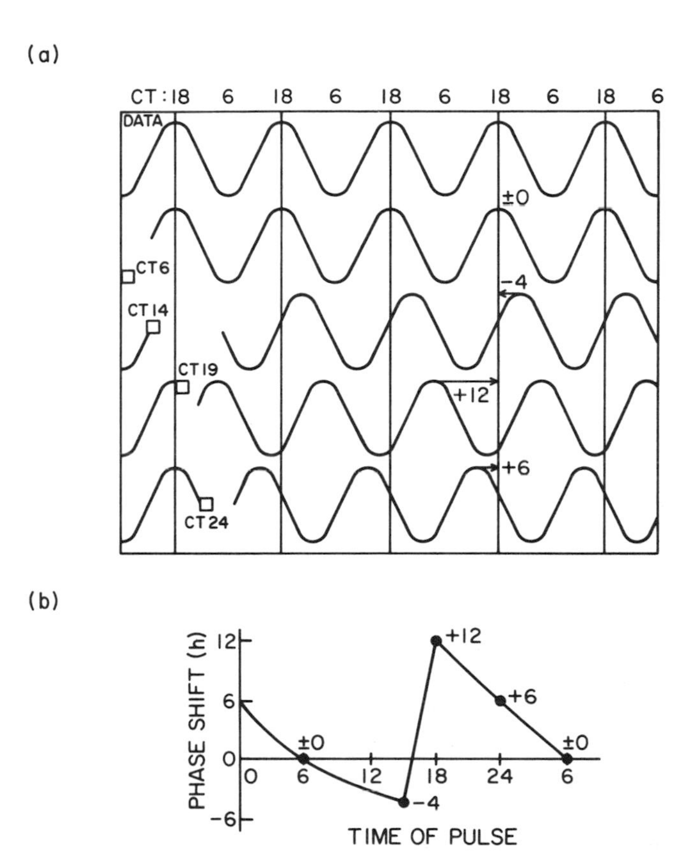

Fig. 2.3. Diagram showing the way in which a phase-response curve (PRC) is obtained. (a) Plotting a phase-response curve. The raw data from untreated and treated samples plotted as amplitude against time. (b) The same data plotted as a PRC, with phase advances positive (+) and delays, negative (−), relative to the untreated control.

ing, respectively, to the light or dark part of a light-dark environmental cycle. The meaning is the same as "day phase" and "night phase."

Entrainment The coupling of the period of one rhythm to that of another cycle of about the same length, for example, the setting of a circadian rhythm to exactly 24 h by the day–night cycle.

frequency demultiplication—entrainment to a long period by a cycle which is a submultiple of that period.

skeleton photoperiod—an entraining light–dark cycle with short light periods marking the beginning of the normal day.

transient—a cycle of an abnormal length immediately following entrainment to a new cycle or a phase shift. More than one transient may be observed before the normal period is reestablished.

Zeitgeber or synchronizer An environmental signal that can entrain a rhythm.

Biological clock or oscillator The mechanism postulated to impart time information to the processes showing rhythmicity. This mechanism is now considered to be intracellular.

chronon—a polycistronic strand of DNA, the transcription of which has been postulated to measure an interval of about 24 h, i.e., act as the circadian oscillator.

Circadian pacemaker The master oscillator in a multicellular organism that coordinates the period and phase of all cellular circadian rhythms. In mammals, the circadian pacemaker is thought to be localized in the superchiasmatic nucleus (SCN) of the hypothalamus in the brain.

desynchronization—a change in the phase relationship between two rhythms in an organism.

Cosinor analysis Data fitted to cosine of desired time dimension, for example 24 h in analysis of a circadian rhythm, by a least-squares procedure. The result is plotted as a circle of 360°, on which the phase angle of the maximum of the rhythm, the "acrophase," is shown by the angular direction of a vector, the amplitude is represented by the length of the vector, and the 95% confidence limits of both are shown as an oval around the tip of the vector (Fig. 2.2).

Photoperiodism Seasonal behavior of an organism in response to the relative duration of day and night in the environment.

Hourglass timer A mechanism capable of timing only one single time period, as is an hourglass, thus not a cycle.

Additional abbreviations used in describing rhythms:

CT Circadian time.

LD A light–dark cycle, for example, L10:D14, a light–dark cycle in which the light period is 10 h and the dark period 14 h.

LL Constant light.

DD Constant darkness.

REFERENCES

Aschoff, J., Klotter, K., and Wever, R. (1965). Circadian vocabulary. *In* "Circadian Clocks" (J. Aschoff, ed.), pp. x–xix. North-Holland Pub., Amsterdam.

Palmer, J. D. (1976). "An Introduction to Biological Rhythms,"Glossary, pp. 363–366. Academic Press, New York.

3

Rhythms That Match Environmental Periodicities: Day and Night

From its beginning, the earth has been in rotation, and the alternation of day and night—the result of this rotation—has dominated the environment of this planet. All living things have evolved under this inviolate periodicity. It is not surprising that they are indelibly imprinted with its frequency. In fact, rhythms with a period close to 24 h come to light everywhere in the biological domain. So common are they that some scientists believe them to be present in every living eukaryote. Although circadian rhythms have not been found in every organism where they have been sought, the possibility remains that we have not examined the right processes.

Of particular interest in this respect is the question whether or not prokayotes have a circadian clock like that in eukaryotes. Some prokaryotes show periodicity in light–dark cycles, but this periodicity does not persist for even one cycle in constant conditions. A number of prokaryotes have been tested for rhythms, but it is difficult to assess how many, since negative results are not usually published. For many years, no circadian rhythms were reported in any prokaryote, eubacterium, archebacterium, or cyanobacterium. However, recently a circadian rhythm in nitrogen fixation in *Synechococcus*, a red cyanobacterium has been reported (Grobbelaar *et al.*, 1986). In a light–dark cycle, fixation of dinitrogen occurred only at night. In constant light, this rhythm persisted for at least 4 cycles. That N_2 fixation should be limited to a time when photosynthesis is not functioning has obvious survival value, since the nitrogenase enzyme complex is inactivated by oxygen. A search of other cyanobacteria that fix nitrogen may yield further examples of circadian clocks in cyanobacteria, especially those without heterocysts. The presence of circadian rhythms in prokaryotes as in eukaryotes is of interest in considering the mechanism of the circadian clock.

Rhythms in which the period is on the order of 24 h were first called diurnal rhythms. Some investigators felt, however, that this might have been a poor choice because "diurnal" is often used to denote the daytime as opposed to nocturnal, at night. To avoid ambiguity, Halberg suggested that these rhythms be called "circadian," from the Latin *circa*, (about) and *diem* (day). This term is appropriate because it stresses one of the most important properties of such rhythms, that the period is often not exactly 24 h under constant conditions. It has come into general usage in spite of a slight awkwardness arising from the similarity between circadian and cicada.

Many circadian rhythms have now been investigated, and the list continues to grow daily. It includes organisms of many levels of complexity: single-celled flagellates like *Euglena, Gonyaulax,* and *Paramecium*, larger algae like *Oedogonium* and *Acetabularia*, molds, Crustacea, many insects, angiosperms, birds, and mammals, including humans. The list of circadian rhythms known in plants would cover many pages, too long to include here. I shall have to limit my discussion to the best known of these, to serve as examples and provide sources of information on others in Table 3.1. Some comparisons with circadian rhythms in animals will be included. However, only circadian rhythms that qualify by the criteria of persistent rhythmicity, with a period close to 24 h under an environment without day–

TABLE 3.1
More Circadian Rhythms in Plants and Where to Find Out About Them

Organism	Rhythm	Reference
Angiosperms		
Onion	UV resistance	Biebl and Kartusch (1973)
Banana leaves	Stomatal opening	Brun (1962)
Dry onion seeds	Gas uptake	Bryant (1973)
Helianthus	Root pressure	Grossenbacher (1939)
Orchid flowers	CO_2 production	Hew *et al.* (1978)
Tomato	Stem growth	Ibrahim *et al.* (1981)
Pea	Expression of mRNAs	Kloppstech (1985)
Chenopodium rubrum	Stem elongation	Lecharny and Wagner (1984)
Pea, soybean, bean	Electron transport water to MV[a]	Lonergan (1981)
Tobacco	Stomatal resistance and root exudation	Macdowall (1964)
Spinach leaves	Membrane potential	Novak and Greppin (1979)
Cestrum nocturnum	Flower opening and odor production	Overland (1960)
Trifolium repens	Sleep movements	Scott and Gulline (1972)
Vicia faba	Opening of stomata	Stålfelt (1965)
Chenopodium rubrum	Energy charge	Wagner *et al.* (1974)
Algae		
Chaetomorpha	Heat resistance	Biebl (1969)
Ulva	Chloroplast migration	Britz and Briggs (1976; 1983)
Chlamydomonas	Phototaxis, cell division	Bruce (1970)
Oedogonium	Spore release	Bühnemann (1955a)
Phytoplankton	Cell division	Chisholm and Brand (1981)
Symbiodinium	Motility	Fitt *et al.* (1981)
Gyrodinium dorsum	Induction of phototaxis	Forward and Davenport (1970)
Chlorella	Photosynthesis	Hesse (1972)
Caulerpa	Photosynthesis	Hohman (1972)
Spatoglossum	Photosynthesis	Kageyama *et al.* (1979)
Symbiodinium	Motility	Lerch and Cook (1984)

(continued)

TABLE 3.1 *(Continued)*

Organism	Rhythm	Reference
Dictyota	Chromatophore movement	Nultsch *et al.* (1984)
Bryopsis	Photosynthesis	Okada *et al.* (1978)
Porphyra yezoensis	Cell size	Oohusa (1980)
Skeletonema	Growth	Ostgaard and Jensen (1982)
Oedogonium	Spore release	Ruddat (1961)
Chlamydomonas	Stickiness to glass	Straley and Bruce (1979)
Griffithsia	Cell division and elongation	Waaland and Cleland (1972)

[a]MV = methyl viologeu

night fluctuations and resettability by light pulses, will be considered (see Chapter 2). Since measurements of rhythmic phenomena must be repeated frequently and over a number of cycles, it is a great advantage to be able to make these measurements automatically. Thus, where possible I have noted methods of automatic recording for the rhythmic organisms discussed.

CIRCADIAN RHYTHMS IN ANGIOSPERMS

Leaf Sleep Movements in Legumes

The leaf sleep rhythms, so common in legumes, are not only the first but one of the most thoroughly studied of circadian rhythms in Angiosperms. The angle that the leaf blade makes with the petiole is easy to measure with a protractor or with a simple lever attached to the leaf tip and a kymograph (Fig. 3.1). Figure 3.2 is an example of Bünning's recordings of the leaf movements of *Phaseolus* in constant light made by this method. Rhythmic sleep movements in the leaves of beans and other legumes make fine experimental material for a laboratory exercise to introduce students to this interesting phenomenon

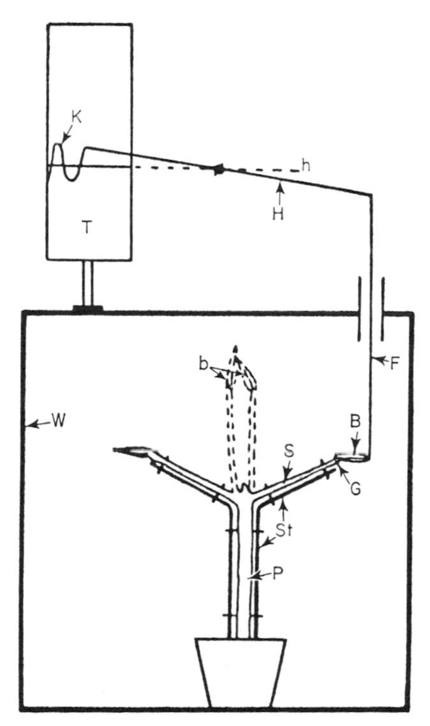

Fig. 3.1. Bünning's apparatus for recording the sleep movements of *Phaseolus multiflorus*. Movement of all parts of the plant except the leaf blade are prevented by restraints. A lever (H) is attached to the blade by a thread (F), and carries a pen at the other end which writes on a rotating kymograph. The plant is enclosed in a box in which temperature and light can be controlled (Bünning, 1931).

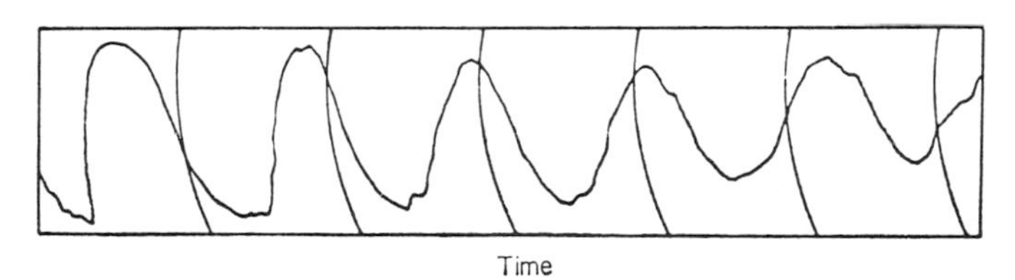

Fig. 3.2. A record of the leaf movements of the primary leaf of *Phaseolus multiflorus* in continuous light and 21 – 22° C. Vertical lines have been drawn every 24 h. (Bünning and Tazawa, 1957).

(Koukkari *et al.*, 1974). Recently more sophistocated measuring devices using light detectors and computers have been devised (Chen *et al.*, 1984). The rhythm in leaves of *Phaseolus* in constant light has been followed for as many as 28 days by time-lapse photography, another method for making measurements automatically (Hoshizake and Hamner, 1964).

Not only *Phaseolus*, but also *Trifolium* and the tree species *Albizzia* and *Samanea*, have yielded interesting insights into the nature of rhythmicity. In the absence of daily light–dark cycles, the leaflets of these legumes continue to show sleep movements just as bean leaves do. *Samanea* and *Albizzia* are particularly nice to work with because their leaves and pulvini are large and compound, allowing paired replicates from the same leaf (Fig. 3.3). It is possible to measure leaf movements with only the pulvinus and a bit of the petiole, the end of which is dipped into a sugar solution (Simon *et al.*, 1976a). In darkness, the movements die out rather quickly unless sugar is fed through the petiole. In addition to sucrose, *Samanea* requires a daily short exposure to red light to continue rhythmic movements in otherwise constant darkness for more than a few cycles (Simon *et al.*, 1976b).

The site of changes that alter the position of leaves is the pulvinus, a specialized tissue at the stem–petiole junction, which acts like a hinge. It is clear that the leaf movements are brought about by the movement of K^+ into certain cells of the pulvinus, followed by the osmotic uptake of water and the swelling of these cells (Satter *et al.*, 1974). On one side of the pulvinus are the "extensor" cells, which swell to cause the leaflets to open (Fig. 3.4), while on the other side are the "flexor" cells, which swell causing the leaflets to close. At night, extensor cells must shrink, as do flexor cells during the day. Ions must both enter and leave extensor and flexor cells to make possible the cyclic movements of the leaves or leaflets. The movement of potassium ions (K^+) in the pulvinus is more complicated than a simple exchange between flexor and extensor regions (Satter, 1979). Other ions in addition to K^+, in particular H^+ and Cl^- (Iglesias and Satter, 1983; Schrempf and Mayer, 1980), may also change location

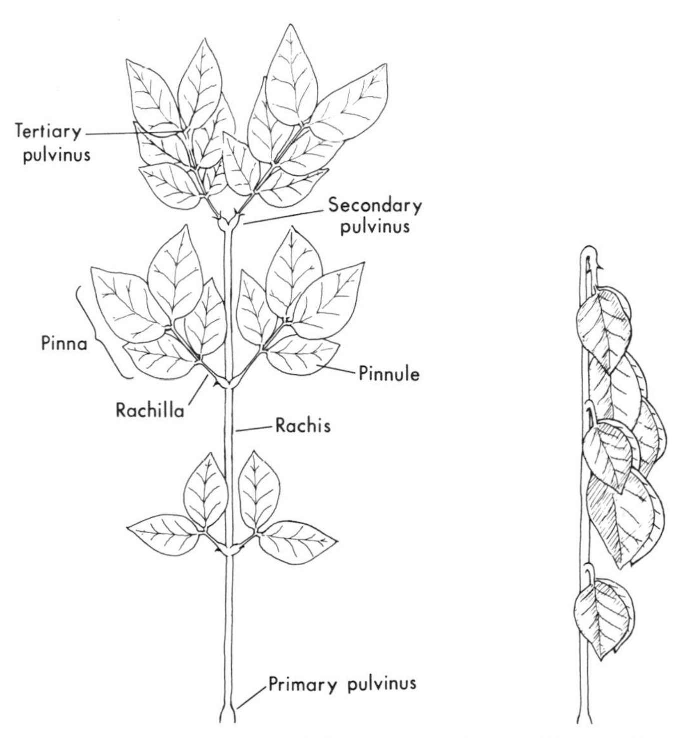

Fig. 3.3. *Samanea saman* leaf in the open-day position (left) and the closed-night position (right) (Satter *et al.*, 1974).

daily. The leaf movements are certainly caused by ion movements. However, whether ion movements in the case of the pulvinar rhythms represent the clock or are overt rhythms, only "hands" of the clock, is not clear at present.

As is typical of circadian rhythms, light can both entrain and reset the leaf movement rhythms. One photoreceptor for these effects of light is phytochrome, since red light pulses reset the rhythm at some times in a circadian cycle and are negated by exposures to far-red light immediately following the red light treatments (Simon *et al.*, 1976b). However, a blue light pulse, if intense enough or long enough, can also change the phase of the rhythm in both *Samanea* and *Albizzia* (Satter *et al.*, 1981). This

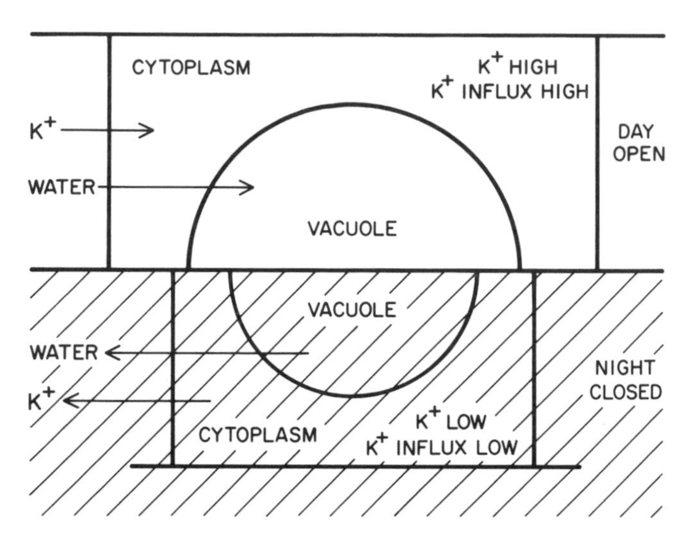

Fig. 3.4. Diagram of a legume pulvinus extensor cell, showing the influx and efflux of potassium ions and water and the attendant changes in volume of the vacuole.

effect is not reversed by a subsequent pulse of far-red light and hence probably is not mediated by phytochrome. Phase-response curves (PRCs) for these light effects on rhythmicity have been determined (Fig. 3.5) and include both delays and advances in phase, as is typical for all circadian rhythms. However, while the shape of the PRC for red light is typical of strong resetting in general, the change from delays to advances in phase takes place at quite a different time in the circadian cycle than in most rhythmic systems—early morning rather than near midnight.

A disadvantage of the leaf sleep movement rhythms for studies designed to extend our knowledge of circadian rhythmicity is that the pulvinus contains many cells besides the extensors and flexors. It has proven difficult to isolate the active cells from the rest, although some progress has been made in this direction by trying to isolate extensor or flexor cells (Gorton and Satter, 1984), but the protoplasts obtained showed evidence of damage during preparation and may not have behaved normally. For example, extensor protoplasts prepared during the day phase contained less potassium than did flexor protoplasts, while the reverse would be expected. Another point to remember is that flexor and extensor cells behave in opposite ways, one swelling while the other shrinks, suggesting that these opposite ion movements are "hands," rather than an intrinsic part of the clock. Further work with isolated protoplasts of extensor and flexor cells, now in progress in Satter's laboratory, may make it possible to answer these questions.

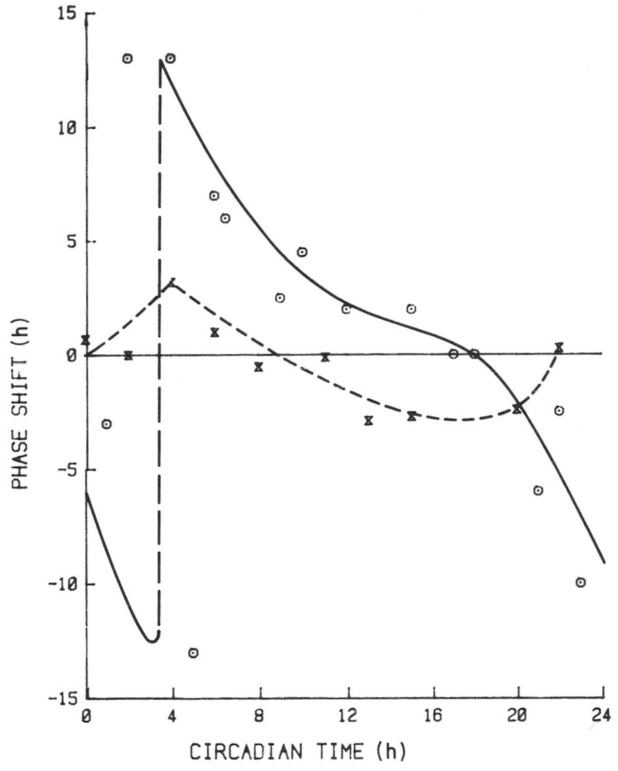

Fig. 3.5. *Samanea saman* phase-response curves to red (circles) and blue (crosses) light. Red light = 1.2×10^{14} quanta cm^{-2} s^{-1} for 5 min. Blue light = 3×10^{14} quanta cm^{-2} s^{-1} for 2 h. (Replotted from Figs. 3 and 4, Satter *et al.*, 1981).

Circadian Rhythms in Succulents

Bünsow (1960), Zimmer (1962), Queiroz (1974), Engelmann (1960), and Engelmann *et al.* (1974, 1978) have investigated another rhythm in plant movement, that of the petals of the flowers of *Kalanchoe blossfeldiana.* These flowers are small, consisting of four petals joined at their bases to form a tubular corolla. At night, the petals move up and in, closing the flower, while during the day they move down and away from each other so that the flower opens. The movement continues in a constant environment in flowers that have been detached from the plant and mounted separately in holders containing a sugar solution. Opening and closing of the flowers can be followed automatically by measuring the changes in the intensity of a light beam directed downward over the flowers (Engelmann *et al.*, 1974). Using this technique, the phase-shifting by light pulses has been studied in detail. Red light at 660 nm is most effective (Fig. 3.6), probably via phytochrome as photoreceptor. The movement of the flowers of *Kalanchoe* is in some respects similar to that of legume leaves, since it is powered by the osmotic swelling of the petal bases, but no organ like the pulvinus is present.

Bryophyllum fedtschenkoi, a succulent plant somewhat similar to *Kalanchoe* in appearance, has been studied by Wilkins in Scotland. Like other succulents, *Bryophyllum* can fix large amounts of CO_2 at night and release it during the day. Using an infrared gas analyzer, Wilkins (1960, 1962) has been able to measure automatically the changes in CO_2 in the gas space around an enclosed leaf (Fig. 3.7). A rhythm in the evolution of CO_2 is initiated by transferring the leaf in the afternoon from light to continuous darkness. The first peak in CO_2 evolution occurs 17–25 h later, depending on the temperature. About three peaks can be observed before the rhythm disappears. This rhythm is not the result of changes in the aperture of the stomata, since Wilkins has been able to demonstrate its presence in mesophyll tissue from which the epidermis has been stripped (Wilkins, 1959) and in tissue culture of mesophyll cells (Wilkins and Holowinsky, 1965). At first it was not clear whether changes in the rate of utilization or evolution of carbon dioxide were re-

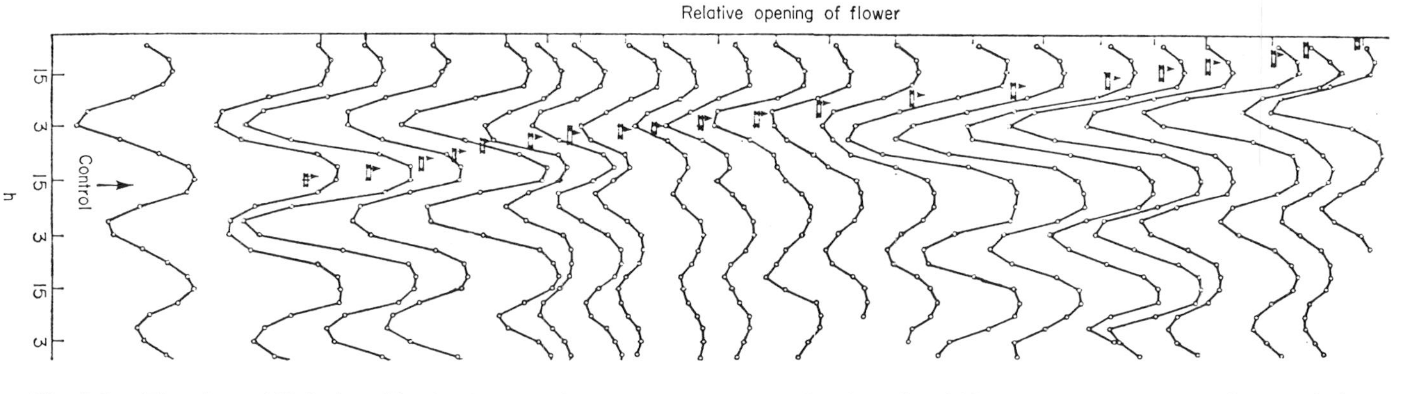

Fig. 3.6. The phase shift induced in the rhythm of petal movement in *Kalanchoe blossfeldiana* by exposure to red light (emission maximum 660 nm) for 2 h, indicated by the bar under each curve. Plants were otherwise in DD (Zimmer, 1962).

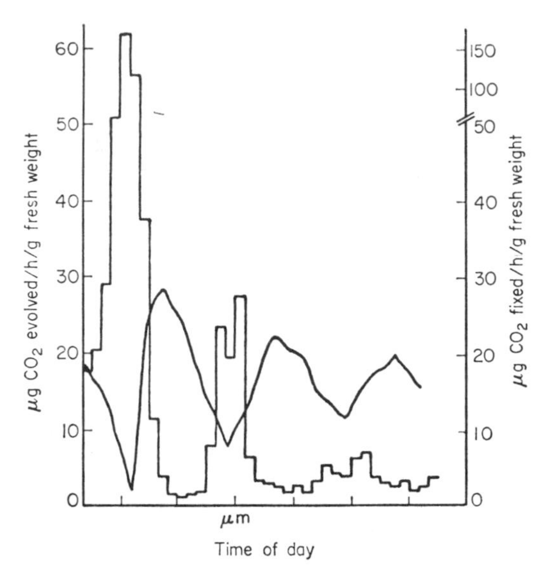

Fig. 3.7. Automatic recording of the CO_2 given off by a leaf of *Bryophyllum fedtschenkoi* in continuous darkness at 26°C (continuous curve) and the measurements of fixation of $^{14}CO_2$ by similar leaves under the same conditions (bar diagram) (Warren and Wilkins, 1961).

sponsible for the rhythm in CO_2. Analysis with $^{14}CO_2$ enabled Warren and Wilkins (1961) to distinguish between these possibilities in favor of differences in the rate of CO_2 fixation. Like other circadian rhythms, the CO_2 rhythm in *Bryophyllum* could be phase-shifted by light (Wilkins, 1973). Only wavelengths between 600 and 700 nm were active, with maximum effectiveness at 660 nm. There was no photoreversal by far-red light, thus phytochrome is not the photoreceptor in this case.

It is both interesting and important to understand rhythms in biochemical terms. In succulents, the uptake of CO_2 at night and its release and refixation by photosynthesis during the day are well known from this point of view, since the process has attracted attention as an example of an adaptation to a dry climate. This process allows the stomata to remain closed during the day and thus conserves water. It is know that CO_2 is fixed by the enzyme, phosphoenolpyruvate carboxylase (PEPcase) (O'Leary, 1982). Oxaloacetate and then malate are formed. Malate is an inhibitor of PEPcase. It is transported into the vacuole where it is stored temporarily, eventually being broken down to CO_2, completing a feedback loop (see Chapter 4). Pulses at high temperature (35 °C) shift the phase of the rhythm in CO_2 uptake. Wilkins (1983) has proposed that this occurs because high temperature opens the gates in the tonoplast, allowing malate to leak out of the vacuole into the cytoplasm. Plant physiologists have been interested in discovering which of these steps gives rise to the circadian rhythm in CO_2 exchange. In *K. blossfeldiana*, as in *Byrophyllum*, the activity of PEPcase changes rhythmically, fixing CO_2 during the night but becoming inactive during the day, possibly by inhibition by malate (Queiroz, 1974; Buchanan-Bollig *et al.*, 1984). This is true also of *B. fedtschenkoi* PEPcase (Wilkerson and Smith, 1976). It is clear from studies using immunoprecipitation with antibodies to PEPcase that this enzyme is not periodically synthesized and broken down but remains at a constant level. In *Bryophyllum*, Nimmo *et al.* (1984) have found that, while the amount of PEPcase remains constant, its sensitivity to inhibition by malate changes in time to the circadian rhythm. They present evidence that this difference in sensitivity is accompanied by a change in phosphorylation of the enzyme, lower sensitivity to inhibition by malate coinciding with phosphorylation, as demonstrated by ^{32}P incorporation *in vivo.* This finding is of particular interest because phosphorylation and dephosphorylation of proteins is known to play an important role in activation and deactivation of enzymes and messenger molecules in animals but has hardly been explored as yet in plants. For a further discussion of this point, see Sweeney (1983).

Circadian Rhythmicity in *Lemna*

The monocots, *Lemna perpusilla* and *L. gibba*, are particularly suitable for physiological studies because they are tiny (only a few millimeters across) and grow naturally floating on water. Furthermore, they can be grown bacteria-free in test tubes and survive in continuous darkness if fed sugar. Fortunately they have a well-defined circadian rhythm in CO_2 output which continues for several cycles in DD, although it damps out in continuous red light. Hillman (1964) has done experiments with *Lemna* showing that the rhythm can be entrained by short pulses of light timed to occur at dawn and dusk just as effectively as by a 12 h light: 12 h dark regime. In addition, the activity of an enzyme phenylalanine ammonia lyase is rhythmic in LL but not in DD (Gordon and Koukkari, 1978). Because *L. perpusilla* is a short-day plant with respect to induction of flowering, this aspect of its rhythmicity will be discussed more fully in Chapter 5.

L. gibba G3 is a long-day duckweed that has been studied by Miyata and Yamamoto (1969), Nakashima (1976), and Goto (1984). As in *L. perpusilla*, this species shows rhythmic CO_2 metabolism. Oxygen uptake continues to oscillate for only two cycles in LL and no rhythm is seen in DD, the opposite situation from *L. perpusilla*. Goto has documented a number of biochemical rhythms involving NAD^+, NADP, NAD kinase, NADP phosphatase, and CA^{2+}-calmodulin. These interesting rhythms are probably not independent, and Goto has suggested that they constitute a self-regulatoring oscillatory loop (Goto, 1984). The glyceraldehyde 3-phosphate dehydrogenases change rhythmically in continuous light, the enzyme in the chloroplasts being highly active at CT 12 when the activity of the enzyme in the cytoplasm is at a minimum. Nakashima (1976) has reported a rhythm in uptake of tritiated uridine, which is temperature-independent between 16 and 30°C in continuous light. This rhythm is the result of a rhythmic change in the activity of RNA polymerase I (Nakashima, 1979).

A rhythm of potassium uptake has been observed in *L. gibba* G3 by Kondo and Tsudzuki (1978). This rhythm persists for 5 days in low, constant light with a period of 25 h at 26 and 30°C, and 24 h at 17°C, so it is temperature-compensated. No K^+ uptake could be observed in the dark. *L. perpusilla*, on the other hand, takes up potassium only in the dark.

CIRCADIAN RHYTHMS IN ALGAE

For studies of rhythmicity, the unicellular algae have many advantages. Circadian rhythms have been found in a number, some of which are very easy to work with since they can be grown with simple equipment in the laboratory under controlled conditions. Thus, material for experiments is always available and it is possible to obtain amounts sufficient for biochemistry. Many algae show circadian rhythms in more than one function, which makes it possible to compare effects from one overt rhythm to another. Table 3.1 gives an idea of the many circadian rhythms that have been studied in algae and where to find more about them. Here we can discuss only a few examples. I have chosen the green algae, *Euglena, Chlamydomonas*, and *Acetabularia*, and the dinoflagellates. With unicellular algae there is no problem identifying the cell responsible for rhythmicity, although the problem still remains of finding the intracellular site of the clock.

Circadian Rhythms in *Euglena*

The first circadian rhythm to be noticed in *Euglena* was that in phototaxis (Pohl, 1948). The cells move toward a light during the day but not at night. Bruce and Pittendrigh (1956) devised a way to record this rhythm automatically, using a test beam that passed down through the cell suspension of *Euglena* held in a Carrell flask and was detected by a photocell beneath. When the cells responded phototactically, they moved into the light beam, which was turned on for 20 min every 2 h, and reduced the light reaching the photocell. The record of the output of the photocell thus showed changes proportional to the number of cells collecting in the test beam during the light exposure (Fig. 3.8). The result showed a clear rhythm with a maximum at midday. Brinkmann (1966) also measured rhythmicity in *Euglena* in a slightly different way, which enabled him to distinguish changes in phototaxis from changes in motility, not possible by the method of Bruce and Pittendrigh. In his experimental arrangement, which resembled that of the Princeton group in principle, the test beam passed through the cell suspension from the side at a point above the bottom of the container. Thus, when the test light was first turned on, the signal detected reflected the number of cells swimming in the medium, while at the end of the 20-min light exposure, phototactic accumulation was registered as the

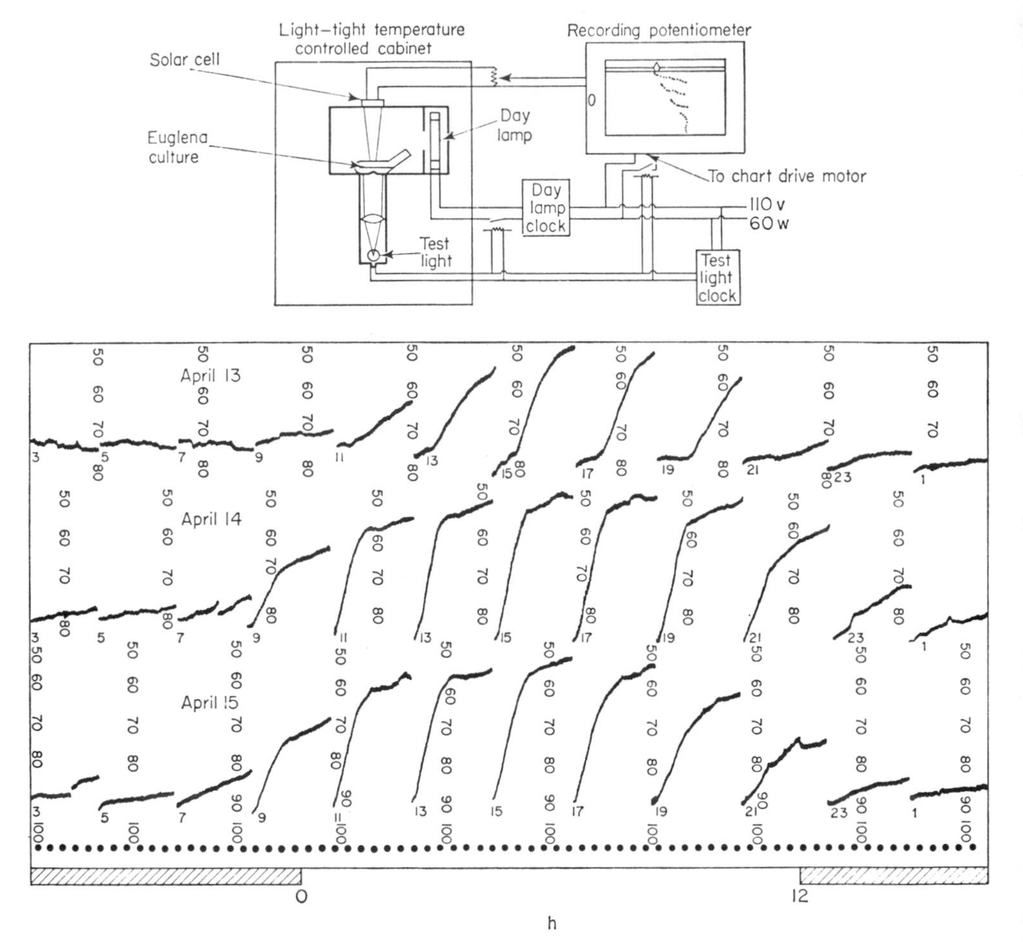

Fig. 3.8. Recordings of the phototactic accumulation of *Euglena gracilis* and a diagram of the apparatus used to make the recording. The test light was turned on for 30 min every 2 h. The 'day' lamp provided LD, 12 : $\overline{12}$ as shown by light and dark bars on the abscissa of the recording. The temperature was 20°C. (Bruce and Pittendrigh, 1956).

decrease in signal during light. He found that, while phototaxis does vary, the largest part of the cyclic measurements comes from changes in motility: at night most of the cells are resting on the bottom of the vessel, while during the day they are evenly distributed at the beginning of the test light exposure.

As in a number of other algae, the cell division in *Euglena* occurs at night in nature. Edmunds and his associates (Edmunds, 1966; Terry and Edmunds, 1970; Edmunds *et al.*, 1982; Malinowski *et al.*, 1985) have studied this phenomenon extensively and have shown that the timing of cell division is a circadian phenomenon, since it continues under conditions that do not give information concerning time of day, such as random or short light–dark cycles (Edmunds and Funch, 1969a, b and Fig. 3.9). The period under these conditions is usually not exactly 24 h and varies only slightly with temperature.

Photosynthesis in *Euglena* can oscillate for as much as three cycles in constant light (Lonergan and Sargent, 1979) and in short light–dark cycles (Laval-Martin *et al.*, 1979). Electron flow through both photosystems, measured as oxygen uptake in the presence of methyl viologen either in whole cells or in isolated chloroplasts, reflects the rhythm in photosynthesis. However, the rate of electron flow through PS I (DCPIPH/ascorbate to MV) is arrhythmic (Lonergan and Sargent, 1979).

Circadian rhythms in phototaxis, motility, and cell division in *Euglena* cannot be measured in constant darkness. The dark period must be interrupted with short light exposures. A further disadvantage is that the rhythms are lost when *Euglena* is grown on organic medium. Still some interesting information has come from the study of rhythms in this green flagellate (see Chapter 4).

Chlamydomonas is another motile green flagellate that has circadian rhythms quite similar to those in *Euglena*, including rhythms in phototaxis, cell division, and motility (Bruce, 1970). In addition, the surface of *Chlamydonomas reinhardi* changes circadianly, giving rise to differences in the cells' stickiness to glass surfaces (Straley and Bruce, 1979). This property makes an easy selection for mutants out of phase with the wild-type cells, since they will be "sticky" at a time when the rest of the population is not. Using this method of selection, Bruce was able to obtain mutants out of phase with the wild type and one mutant that differed in phase and so was maximally phototactic at a different time from the wild type (Bruce, 1972).

Spudich and Sager (1980) also studied cell division in *Chla-*

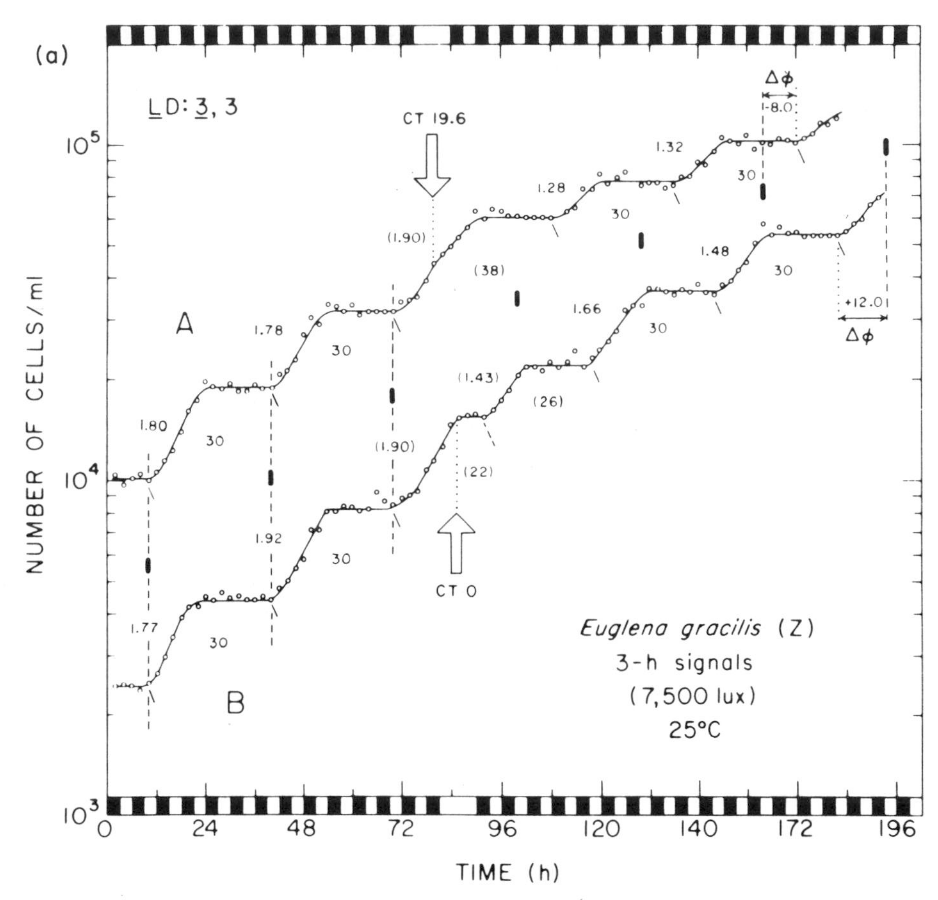

Fig. 3.9a. The rhythm in cell division in *Euglena gracilis* in LD 3 : 3, 25 °C. After at least three cycles of the rhythm with a 30-h period, one culture (A) was exposed to 3 h light at CT 19.6 and the other (B) at CT 24. The phase relative to that before the light pulse was measured after several cycles and considered to be the phase-shift (Edmunds *et al.*, 1982).

mydomonas. They found that without a 4-h light exposure during G_1 of the cell division cycle, this alga would not complete the cell cycle. This finding is certainly reminiscent of the finding that *Euglena* would not continue to divide in DD without a short, intermittent exposure to light.

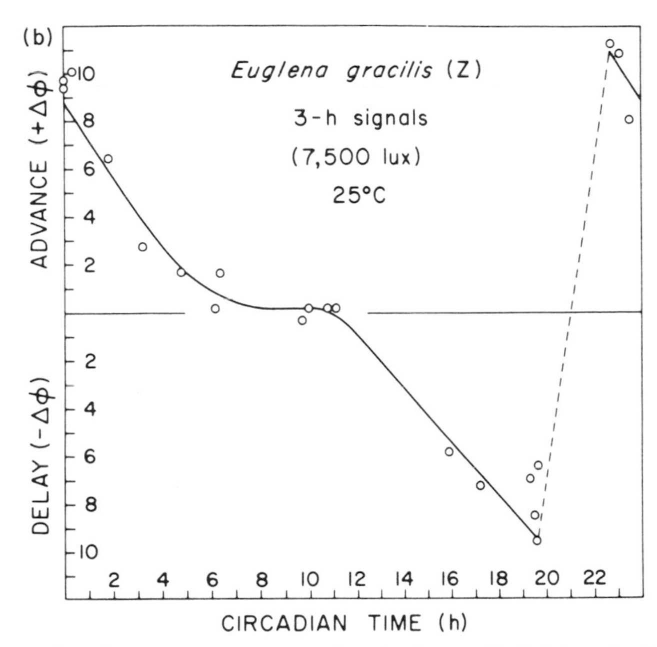

Fig. 3.9b. The phase-response curve for *Euglena* cell division rhythm obtained from experiments like that shown in Fig. 3.9a (Edmunds *et al.*, 1982).

Circadian Rhythms in *Acetabularia*

Acetabularia is a most unusual alga because it is large for a unicell, but particularly because its nucleus, which occupies the holdfast region of the cell at most stages of the life cycle, can be removed easily; the cell can survive for a number of weeks without a nucleus, carrying out most of its everyday functions, including photosynthesis and the formation of a cap. Thus, it is an ideal organism for answering the question of whether or not the nucleus is necessary for the running of a circadian rhythm, provided that a circadian rhythm can be found in this organism. Fortunately, *Acetabularia* has a clear circadian rhythm in photosynthesis. This rhythm has been shown by five different laboratories not to require the presence of the nucleus (Sweeney and Haxo, 1961; Richter, 1963; Schweiger *et al.*, 1964; Vanden

Driessche, 1966a; Terborgh and McLeod, 1967). The rhythm in photosynthesis has been found in all the species of *Acetabularia* in which it has been looked for, including *A. mediterranea, A. crenulata*, and *A. major*. Automatic recording of this rhythm can be made with oxygen electrodes that feed data into a computer. Even small fragments of *Acetabularia* cells show a circadian rhythm in oxygen evolution (Mergenhagen and Schweiger, 1975). Only the cap containing reproductive cysts is arrhythmic.

Other circadian rhythms can also be demonstrated in *Acetabularia*. If electrodes are placed near the tip and the base of a plant, a potential difference can be detected (Novak and Sironval, 1976). The magnitude of this potential varies with time of day and this change occurs in constant light (Broda and Schweiger, 1981). In addition, the chloroplasts move up and down the stem region of the plant, and their movement can be followed by measuring the transmission of the lower part of the stem to light, thus demonstrating a circadian rhythm (Broda *et al.*, 1979). Automatic devices for measuring both these rhythms have been devised in Schweiger's laboratory at the Max Planck Institüt für Zellbiologie (Broda *et al.*, 1979; Broda and Schweiger, 1981). An example of a long series of measurements of these rhythms recorded automatically is the frontispiece of this book. This figure shows that the rhythm in chloroplast movement and electrical potential difference are similar and also that an 8-h dark pulse given to cells otherwise in LL on day 235 changed the phase of both rhythms on subsequent days.

In Schweiger's laboratory, proteins extracted from *Acetabularia* at different times of day in LL are being separated (Leong and Schweiger, 1979). The amounts of several proteins vary with time, one of them a large protein that may be in the chloroplast membrane (Hartwig *et al.*, 1985). Now it must be determined whether these proteins are rhythmic because they are "hands" of the clock, or, more interesting, a part of a circadian oscillator. As we shall see later, inhibitors of protein synthesis shift the phase of the circadian rhythm in oxygen evolution in *Acetabularia*, and also in the eye of *Aplysia*, a sea slug, and in the dinoflagellate *Gonyaulax*. Thus, a protein or several proteins may be important components of the clock.

Vanden Driessche in Brussels has also carried out a long series of experiments on *Acetabularia*. She has found that the shape of the chloroplasts changes in time with the rhythm in photosynthesis, these organelles being longer during the light phase, both in LD and in LL, whether or not the cells are enucleated (Vanden Driessche, 1966a; Vanden Driessche and Hars, 1972a, b). Both the rhythms in oxygen evolution and chloroplast shape are sensitive to actinomycin D in intact cells but *insensitive* to this inhibitor of transcription when enucleated (Vanden Driessche, 1966b). Vanden Driessche also examined the possibility of circadian rhythms in ATP content (Vanden Driessche, 1970) and in incorporation of [^{3}H]uridine into RNA in *Acetabularia* (Vanden Driessche and Bonotto, 1969), but the results were equivocal because circadian differences in those cell components in LL were small and variable. The inhibitor of phosphodiesterase, theophyllin ($2 \times 10^{-3} M$), had no apparent effect on rhythmicity in *Acetabularia* (Minder and Vanden Driessche, 1978), although cAMP was present. Interestingly, in both whole cells and isolated chloroplasts, fatty acids increased by a factor of 3 during the last part of the day in LD (Jerebzoff and Vanden Driessche, 1983). Unfortunately no data are given for plants in LL in this study.

Circadian Rhythms in Dinoflagellates

The dinoflagellates are unicellular organisms of the plankton of both freshwater and saltwater. Many, although not all, species contain chloroplasts and some are bioluminescent. A long time ago, I found by accident that the bioluminescence of the marine dinoflagellate *Gonyaulax polyedra* (Fig. 3.10) is much brighter at night than during the day. Furthermore, this behavior continues for several days in constant darkness (Haxo and Sweeney, 1955; Sweeney and Hastings, 1957). Later, Hastings and I (1958) found that the circadian rhythm in bioluminescence also persists in LL. Under these conditions many cycles can be measured. Because of this, we were able to investigate the effect of different temperatures on the period of the rhythm. The result is shown in Fig. 3.11, which has several interesting features. Notice that the

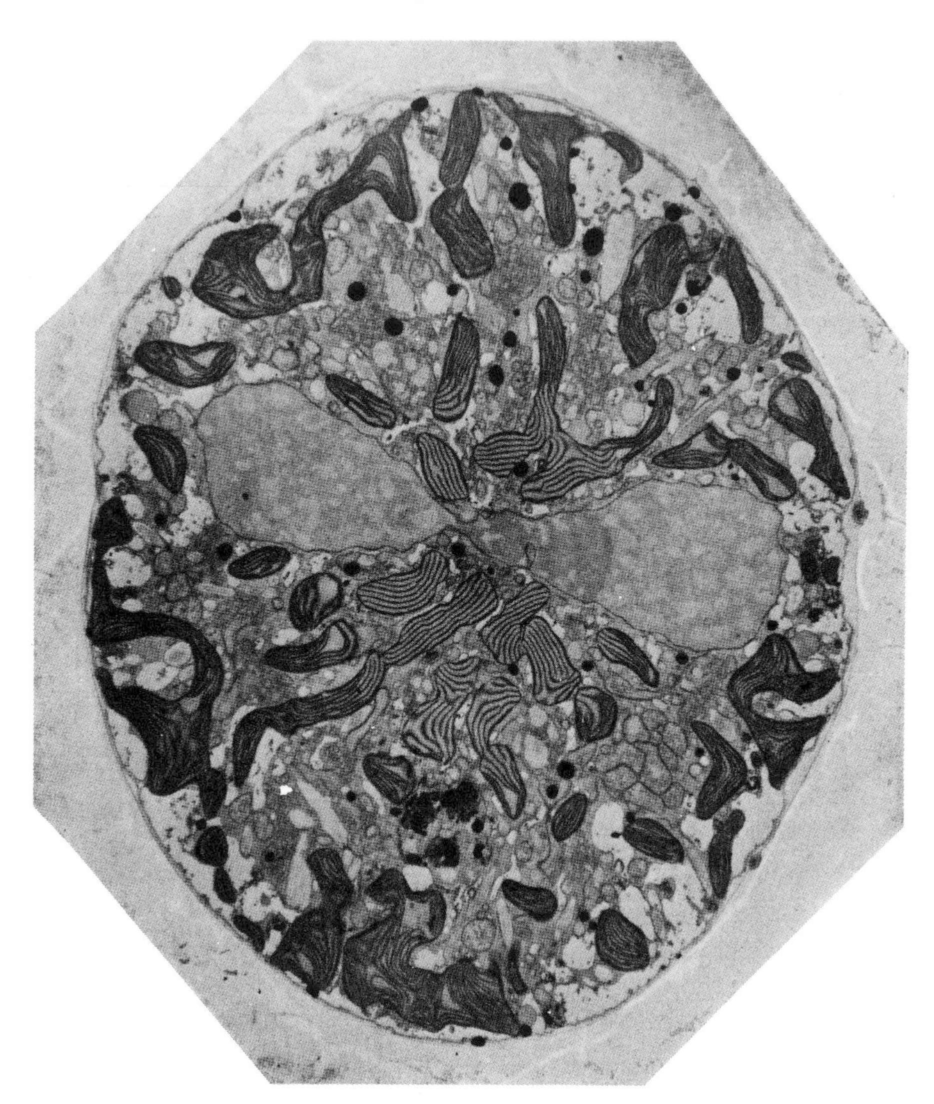

Fig. 3.10. Where is the clock? *Gonyaulax polyedra* in longitudinal section showing the nucleus, chloroplasts, mitochondria, and other structures. Courtesy of B. Bouck.

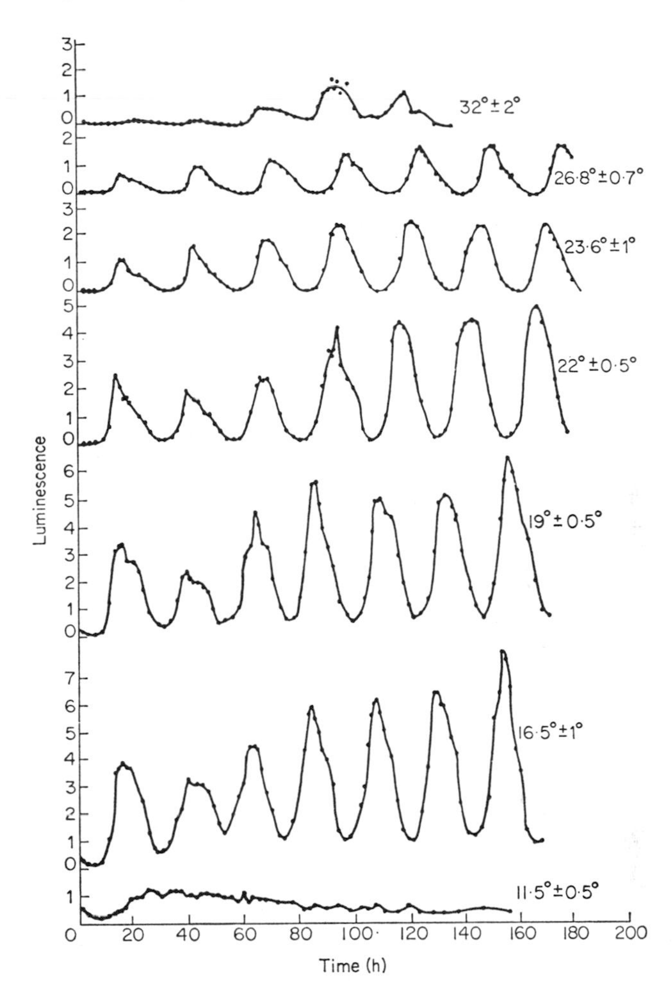

Fig. 3.11. The rhythm in stimulated luminescence in *Gonyaulax polyedra* in LL (1000 lux) at different temperatures °C as indicated on each curve. Note that the period changes a little as the temperature is changed (Hastings and Sweeney, 1957).

period becomes longer at higher temperatures, which is not what one would expect from a chemical reaction, most of which run faster as the temperature is increased. However, this effect is rather small. This experiment is an example of temperature compensation rather than true temperature independence. A similar small effect of temperature on period is seen in many circadian rhythms in both plants and animals. Notice also that the bioluminescence becomes dimmer at higher temperatures and finally is almost extinguished at 30°C. At low temperature, the bioluminescence, which normally requires mechanical stimulation, is emitted spontaneously so that a later stimulation does not induce light. This experiment with *Gonyaulax* was one of the first demonstrations of the remarkably small effect of temperature on the period of a biological clock.

Although at most times a stimulus is required to elicit bioluminescence, there is a time during the latter part of the night when light is observed in the absence of stimulation (Fig. 3.12), cells either occasionally flashing or emitting a faint glow. Thus, the circadian rhythm in bioluminescence can assume three different forms, stimulated emission, spontaneous flashing, and a glow. The fact that no stimulus is needed for light emission has been an advantage for designing automatic devices to measure the rhythm in bioluminescence. The laboratories of Hastings and of Rensing have these "Taylortrons," named for Walter Taylor, who did much of the design and programming of the accompanying computer. Such devices are especially convenient for assaying the effects of different substances on the period or the phase of the bioluminescence rhythm. For the former, the cells are exposed to a drug continuously, for the latter, only for a short time. The results of these studies, discussed in Chapter 4, have given us some insight into the nature of the circadian clock.

If we are ever to understand what directs the circadian rhythms, it is of great importance to know what biochemical changes underlie the rhythms that we usually measure, bioluminescence, for example. With this in mind, Bode extracted both luciferin, the substrate for the reaction by which light is emitted (Hastings and Bode, 1962), and luciferase, the enzyme

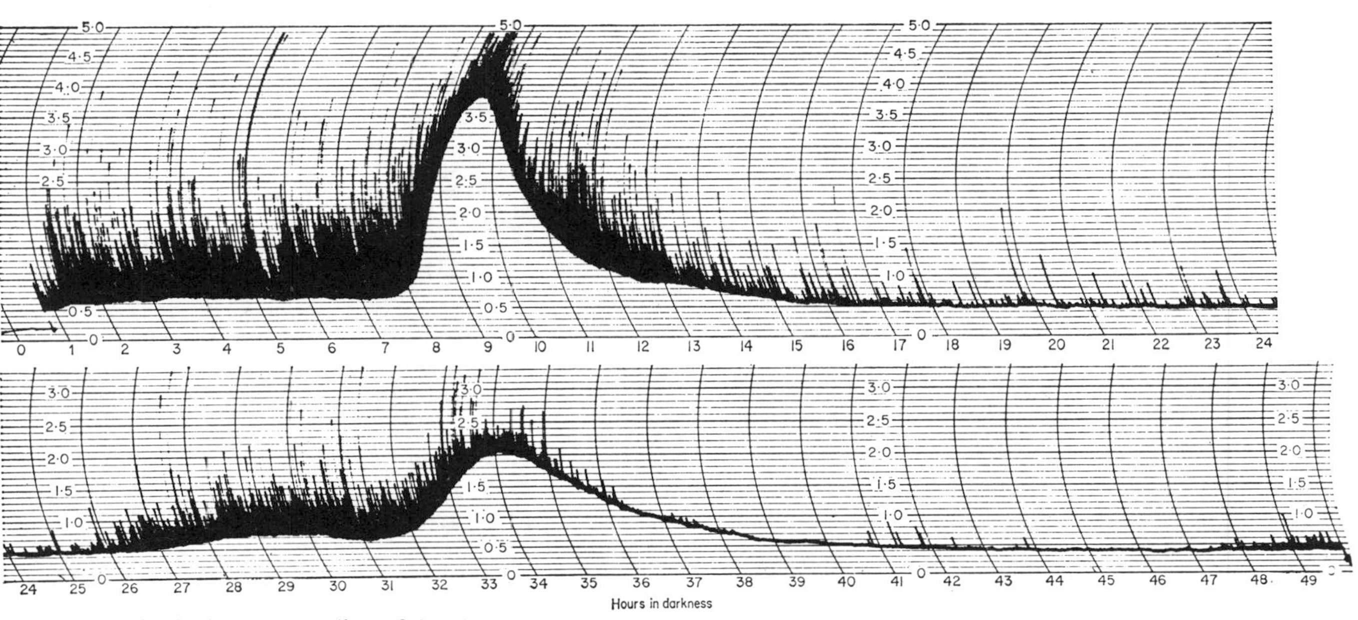

Fig. 3.12. A recording of the glow rhythm in *Gonyaulax polyedra* in DD (B. M. Sweeney, unpublished).

responsible (Bode *et al.*, 1963). He found that the activity of both varied over the day and night, even when the cells were in constant light. Using an antibody to the luciferase extracted from *Gonyaulax*, Dunlap has shown that the *amount* of luciferase in the cells varies with the circadian cycle (Dunlap and Hastings, 1981) and this has been confirmed in additional experiments (Johnson *et al.*, 1984). Yet the biochemistry of the luminescent reaction is certainly not the clock, since bioluminescence can be stimulated to exhaustion without affecting the subsequent timing of maxima nor indeed changing the time of other rhythms in *Gonyaulax*.

One might wonder whether cells in a culture can talk to each other, by their flashes or otherwise, communicating information about the time. Experiments in which cultures from different light schedules, hence different in circadian time, were mixed did not show any evident mutual interaction (Hastings and Sweeney, 1958). Recent experiments using the glow rhythm recorded with the Taylortron suggested that a very small effect of one culture on another might be detected after 5 or so days in LL (Halberg *et al.*, 1985). However, this result requires confirmation.

Gonyaulax has been the subject of many experiments in addition to those concerned with its bioluminescence. Its photosynthesis is now known to be rhythmic, with a maximum near midday and the property of continuing for a long time in LL like bioluminscence (Sweeney, 1960, 1979; Hastings *et al.*, 1961; Prézelin *et al.*, 1977a, b). The biochemistry of photosynthesis is now well understood, making it possible to ask experimentally which partial reactions are responsible for rhythmicity. Experiments with cell-free preparations of *Gonyaulax* show that rhythmic fluctuations in activity can only be seen in photosystem II and not in photosystem I (Samuelsson *et al.*, 1983). The activity of ribulose bisphosphate carboxylase assayed in extracts did not show circadian variations (Bush and Sweeney, 1972).

Cell division, too, shows evidence of clock control, cells dividing exclusively at the end of the night phase in a 12 : 12 light–dark cycle and after transfer to LL (Sweeney and Hastings,

1958). The time that the cells divide is apparently set by the beginning of darkness, since cells kept on a schedule of 8 h light : 8 h darkness divide about 4 h after the lights turn *on*, i.e., 12 h after lights off (Hastings and Sweeney, 1964). In fact, it is clear that the timing of the cells is not learned. When they are returned to constant light after a whole year in a 7 : 7 LD cycle, they immediately continue their circadian rhythm with about a 24-h, rather than a 14-h period.

Another rhythmic process in *Gonyaulax* can be detected in freeze-fracture preparations where the cells are quick-frozen, fractured under vacuum, and replicated with platinum and carbon. This technique allows us to see particles in the plane of membranes, the intramembrane protein complexes. *Gonyaulax* has a large flattened vacuole completely encircling the cell just under the cytoplasmic membrane. One face of this membrane is seen to contain more of these particles at night in LL than during the day phase (Fig. 3.13), and some of the particles are larger (Sweeney, 1976). On the other hand (Sweeney, 1981a), all the membranes of the chloroplast, including the three membranes of the chloroplast envelope and the thylakoid membranes, contained the same number and size of particles at all the times examined (four times during one cycle in LL). What these findings mean is not yet clear.

The membrane resting potential of dinoflagellates is not easy to measure. *Gonyaulax* is so packed with intracellular structures that electrodes immediately become plugged when inserted. However, it is possible to measure the relative resting potential indirectly by following the fluorescence of a cyanine dye. This procedure was tried with *Gonyaulax* by Adamich *et al.* (1976) and a circadian rhythm could be demonstrated by this means.

Another provocative circadian rhythm, that in the amount and nature of cellular RNAs, was documented by Brigitte Walz (Walz *et al.*, 1983). As discussed in Chapter 4, the failure of actinomycin, an inhibitor of transcription, to reset the phase of the rhythm in bioluminescence suggests that the RNA rhythm is not part of the clock oscillator.

Other dinoflagellates besides *Gonyaulax* show circadian

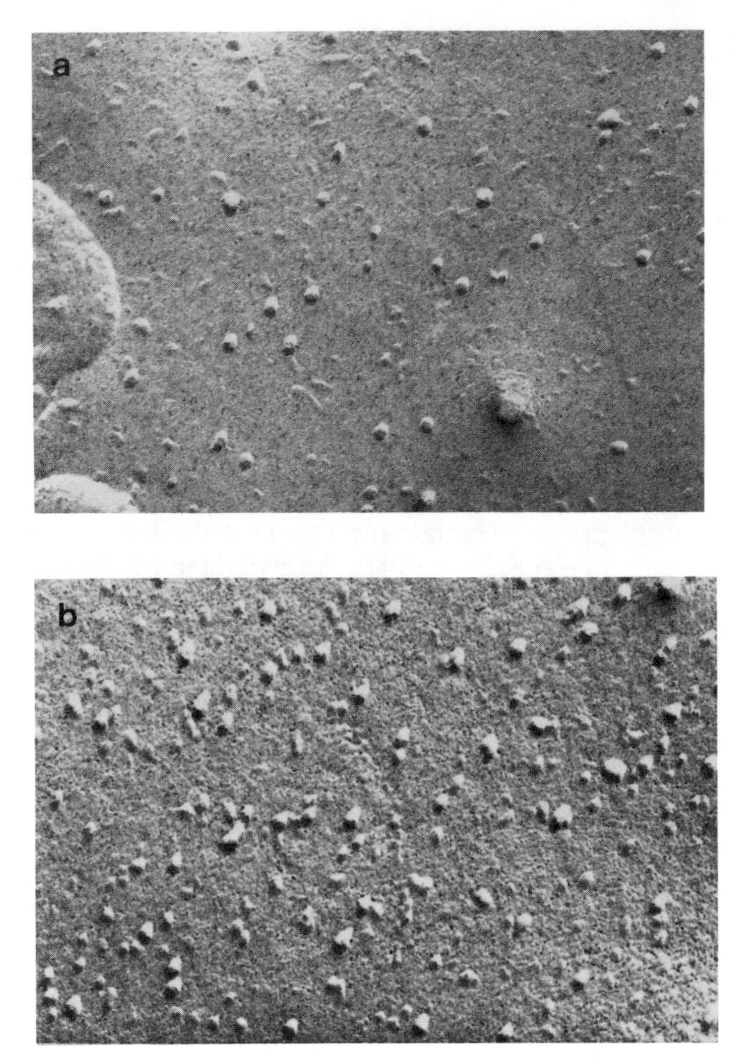

Fig. 3.13. The E_f face of the peripheral vacuole of *Gonyaulax polyedra* in LL, freeze-fractured at CT 06 (a) and CT 18 (b). Note that there are more particles on the membrane during the night phase, some of which are also larger. Magnification, × 170,000. (Rephotographed from Sweeney, 1976).

rhythms and thus also have a circadian clock. The brilliant bioluminescence of *Pyrocystis fusiformis* is clearly brightest at night and during the subjective night in LL or DD (Sweeney, 1981b), a rhythm that is apparent in single cells (Sweeney, 1982). A circadian rhythm in photosynthesis has been demonstrated in *P. fusiformis*, and the nonbioluminescent species, *Ceratium furca* and *Glenodinium* sp. Maxima in oxygen evolution or carbon dioxide uptake in light are found in the middle of the day, rather than at night like bioluminescence. The pattern repeats with a circadian period when the cells are in continuous light.

As in *Gonyaulax*, cell division in a number of other dinoflagellates takes place during the night, the subjective night if the cells are in LL. In some species, for example *Prorocentrum micans*, cell division occurs during the afternoon in nature. Since the cell division rhythm continues in constant conditions, it is under circadian control. Other examples of circadian rhythms in cell division can be found in *Ceratium furca* (Meeson, 1977), *Cachonina illdefina* (Herman and Sweeney, 1976), *P. fusiformis* (Sweeney, 1982), and *P. lunula* (Swift and Taylor, 1967), even in the heterotrophic dinoflagellate, *Noctiluca miliaris* (Uhlig and Sahling, 1982). Interestingly, although brightly bioluminescent, *Noctiluca* lacks a rhythm in bioluminescence.

Cell division is especially easily studied in the large bioluminescent dinoflagellate from subtropical seas, *P. fusiformis* (Fig. 3.14). The generation time of this organism is quite long, 4–6 days, allowing time to follow other stages in the cell cycle in addition to cytokinesis. Helpful for this is the fact that *Pyrocystis* changes morphology several times during one cell division cycle. By making repeated observations of isolated cells, it was possible to detect when one morphological stage changed to the next. It was clear that these changes in morphology, like cell division itself, all occurred at certain defined times and were in fact also controlled by the circadian clock (Sweeney, 1982).

In both *Gonyaulax* and *Pyrocystis*, there are changes in the plastids that show a circadian rhythmicity. During the day the chloroplasts of *Gonyaulax* in the inner part of the cell have

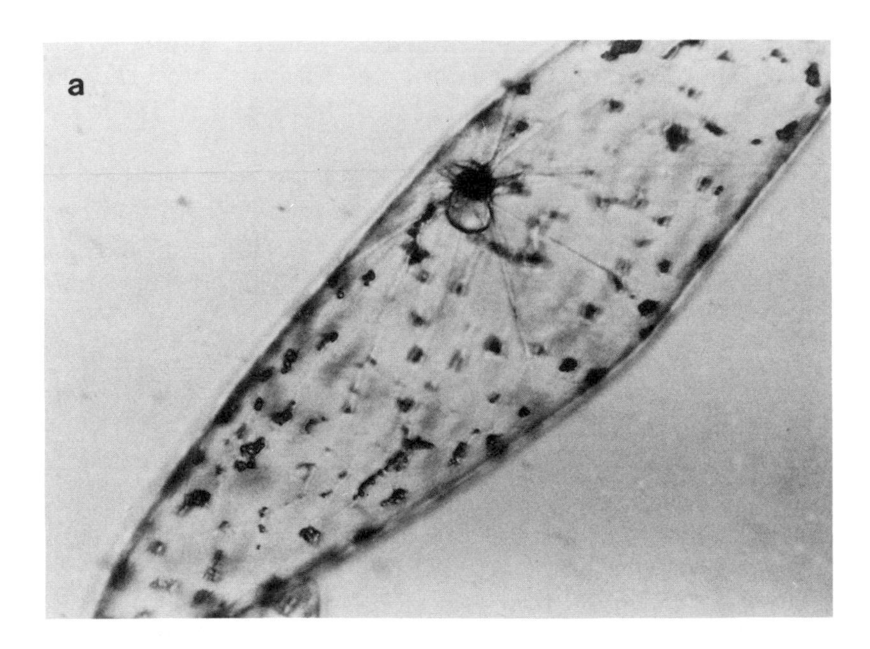

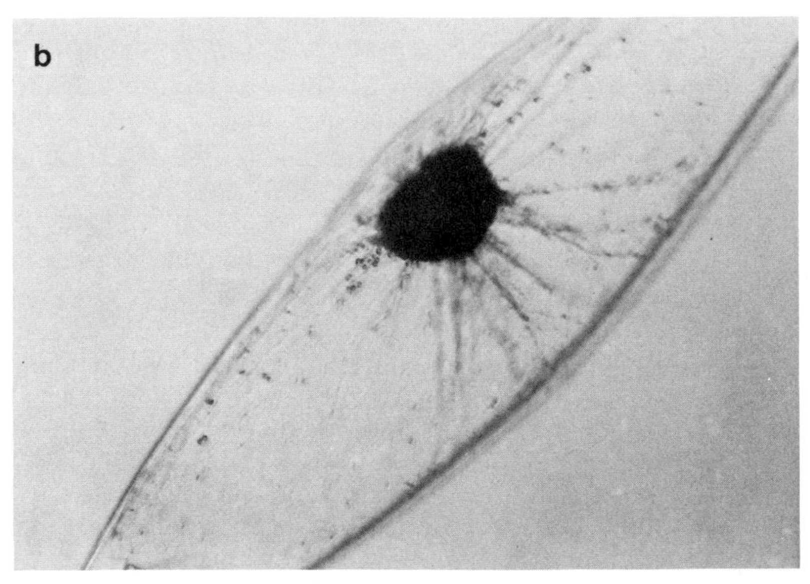

Fig. 3.14. Light micrograph of *Pyrocystis fusiformis* taken (a) during the day phase of the first day after cell division and (b) during the subsequent night phase (B. M. Sweeney, unpublished).

absent at night, both in LD and LL (Herman and Sweeney, 1975). It is possible that peripheral chloroplasts move outward (Rensing *et al.*, 1980), but the difference is hard to see in electron micrographs, impossible without it. On the other hand, the chloroplasts in *P. fusiformis* can be seen in the dissecting scope to move into the center of the cell at night and out again before dawn (Sweeney, 1981b). This movement continues in LL and constitutes a very easily observed circadian rhythm. How they move so far, up to 500 μm within less than 1 h, is now being investigated by a student in my laboratory. The chloroplasts of another species of *Pyrocystis*, *P. lunula*, also move rhythmically in LL (Swift and Taylor, 1967; Topperwein and Hardeland, 1980).

When dinoflagellates symbiotic in *Aptasia tagetes* were cultured in the laboratory, they become motile but only at certain times in a circadian cycle, the peak occurring at CT 04 in DD (Lerch and Cook, 1984). Symbionts isolated from other hosts showed maximum motility at other times characteristic of their host origin (Fitt *et al.*, 1981).

The many rhythmic processes demonstrated in dinoflagellates make this group good subjects for the study of circadian rhythmicity. Most of the rhythms we have discussed are almost certainly "hands," rather than parts of the clock itself. Dinoflagellates as experimental organisms with which to clarify the nature of the circadian clock have certain disadvantages, principally the difficulty of inducing sexuality in many of this group. Hence, little is known about their genetics. Organelles are also difficult to isolate from members of this group.

CONCLUSIONS ABOUT THE GENERAL PROPERTIES OF CIRCADIAN RHYTHMS

In the sections above, we have introduced the plant rhythms about which we know the most details. What do they all have in common that can help us understand the phenomenon of circadian rhythmicity? As we have said, in order to be included among the circadian rhythms, the oscillations in various functions must have been shown to be independent of any environmental 24-h changes such as the day–night variations in light and temperature. The rhythms must be endogenous.

Given that the changes we observe are true circadian rhythms, what properties do they possess in common? First, in many different organisms, including invertebrate and vertebrate animals as well as the plants we have described, the natural period of the rhythms in an unchanging environment is not exactly 24 h, in fact, is not a constant. The period changes with the temperature and irradiance, varying from a little less to a little more than 24 h. Since this is true for all circadian rhythms, how are they synchronized with day and night? Under natural light or a 12 : 12 light–dark cycle in the laboratory, all rhythms show periods that are exactly 24 h. They are entrained by the cycles of light and temperature in the environment. Rhythms in plants can often be entrained to unnatural light–dark cycles if the timing of the cycles is not too different from 12 h light : 12 h darkness, as for example 7 h light : 7 h darkness in *Gonyaulax*. How does this come about?

Sensitivity to Light

A universal feature of the circadian rhythms is their sensitivity to light. This is clearly seen in the entrainment of rhythms to the light–dark cycle of day and night. Entrainment can be understood as the sum of resetting of the phase of the rhythms by single short light pulses, delaying in the early night and advancing after midnight, as manifest in the PRCs, very similar in shape in all rhythmic organisms (Fig. 3.15). This PRC seems to be the most accurate reflection available of the phase of the oscillation underlying the overt rhythms. Phase delays by light in the early evening cause behaviors and processes appropriate to darkness to await the beginning of night, while advances in phase by light in the early morning prevent behavior appropriate to darkness from lasting into daylight. Thus, the characteristics of PRCs to light pulses have selection value. Experiments where two light pulses were given in close succession led to the conclusion that

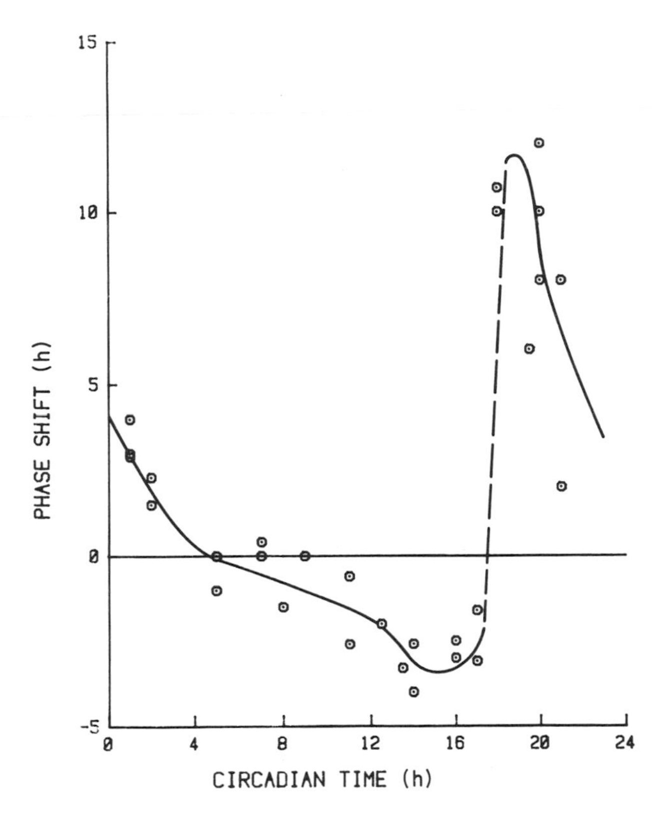

Fig. 3.15. Phase-response curve for the bioluminescence rhythm in *Gonyaulax polyedra*, redrawn from Fig. 3.17 in "Rhythmic Phenomena in Plants", First Edition, to plot phase advances as positive, phase delays as negative.

resetting takes place very quickly, at least when the resetting light is bright. The effect of the second light pulse already is that of the rhythm shifted by the first pulse.

The absence of light during the day part of the circadian cycle also resets phase, with a PRC like that to light but displaced by about 12 h, with delaying phase-shifts occurring in the morning, and advancing phase-shifts occurring in the afternoon (Karakashian and Schweiger, 1976).

While all circadian rhythms that have been examined in this

respect are phase-shifted by light pulses, the photoreceptor which senses the light must be different in different organisms. Available photoreceptors seem to have been pressed into the service of the circadian clock, as shown by the action spectra for phase-shifting that differ from organism to organism. For example, phytochrome acts as photoreceptor for the circadian responses of legumes, but does not have this role in succulents and dinoflagellates, in which there is no evidence for reversal by far-red wavelengths. In dinoflagellates, the action spectrum for phase-shifting shows peaks in both red and blue light, but no reversal by far-red light (Fig. 3.16). Blue light receptors are active in resetting other rhythms, such as that in eclosion in *Drosophila* (Klemm and Ninnemann, 1976) and the activity rhythm of the golden hamster (Takahashi *et al.*, 1984).

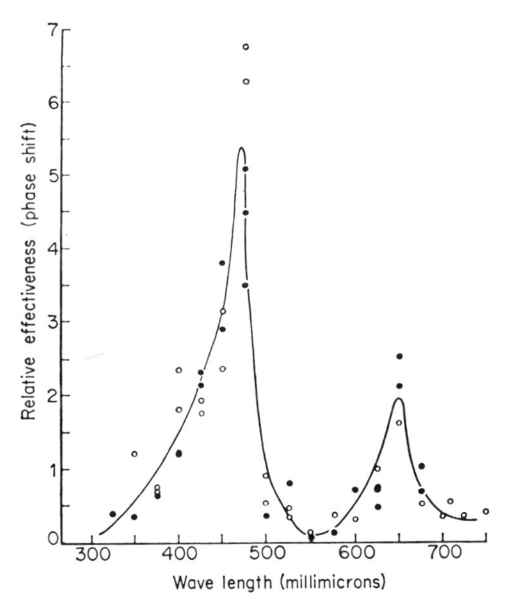

Fig. 3.16. The action spectrum for resetting phase in the rhythm of stimulated luminescence in *Gonyaulax polyedra*. Cells otherwise in DD were exposed to light of different wave lengths for 3 h at the middle of the night phase (Hastings and Sweeney, 1960).

Temperature Sensitivity

As we have seen in *Gonyaulax* (Fig. 3.11), the period of most circadian rhythms is remarkably well temperature-compensated. Usually, however, there is a slight temperature effect (Table 3.2). Sometimes the period becomes slightly *longer* as the temperature is raised, contrary to what one might expect and one of the things that suggests temperature compensation rather than temperature independence.

While light is the dominant phase-shifting stimulus, in a constant environment with respect to light, temperature pulses can reset circadian rhythms (Fig. 3.17). As we have seen, the effect of changing the environmental temperature for times long relative to 24 h is small but almost always measurable. However, the result of pulses of higher or lower temperature can be a large phase-shift.

SUMMARY

The examples described in this chapter and many others make it clear that circadian rhythmicity is common in plants, as it is

TABLE 3.2
Temperature Dependence in Various Plant Circadian Rhythms

Plant	Rhythm	Q_{10}[a]	Reference
Euglena	Phototaxis in DD	1.01–1.1	Bruce and Pittendrigh (1956)
Gonyaulax	Luminescence in LL	0.85	Hastings and Sweeney (1958)
	Cell Division in LL	0.85	Sweeney and Hastings (1958)
Helianthus	Exudation	1.1	Grossenbacher (1939)
Oedogonium	Sporulation in LL	0.8	Bühnemann (1955b)
Phaseolus	Sleep Movements	1.3	Bünning (1931)
		1.01	Leinweber (1961)

[a] $Q_{10} = \frac{\text{frequency at T}}{\text{frequency at T}} - 10$ or $\frac{1/\text{period at T}}{1/\text{period at T}} - 10$

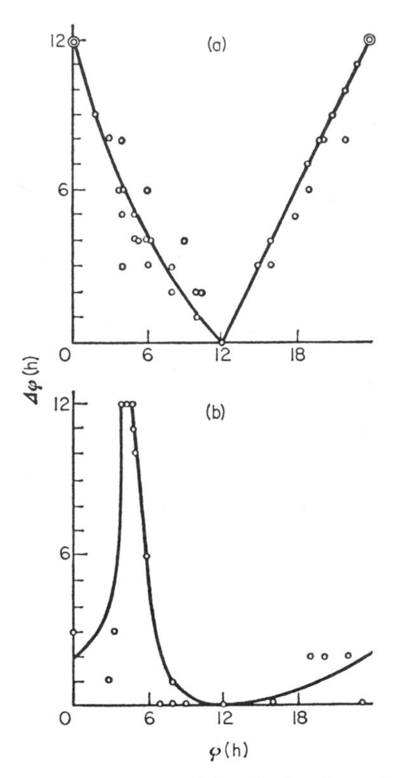

Fig. 3.17. Phase-response curves of the rhythm in mobility and phototaxis in autotrophic cultures of *Euglena gracilis* to raising (curve a) or lowering (curve b) the temperature by 5 °C in the temperature range where the period is unaffected by change in temperature (0 time is taken as the middle of the night phase when mobility is minimal) (Brinkmann, 1966).

also in animals. The properties of rhythms in different organisms, even those widely separated in evolution, are remarkably similar. Circadian rhythms can be observed in biochemistry as well as in behavior. They must have great value in adapting to the day–night cycle of our environment.

REFERENCES

Adamich, M., Laris, P. C., and Sweeney, B. M. (1976). *In vivo* evidence for a circadian rhythm in membranes in *Gonyaulax. Nature (London)* **261**, 583–585.

Biebl, R. (1969). Studien zur Hitzeresistenz der Gezeitenabe *Chaetomorpha cannabina* (Aresch) Kjellm. *Protoplasma* **67**, 451–472.

Biebl, R., and Kartusch, R. (1973). Tagesschwankungen der U.V.-Strahlen-resistenz bei *Allium cepa* L. *Protoplasma* **76**, 227–234.

Bode, V. C., DeSa, R., and Hastings, J. W. (1963). Daily rhythm of luciferin activity in *Gonyaulax polyedra. Science* **141**, 913–915.

Brinkmann, K. (1966). Temperature influsse auf die circadiane Rhythmik von *Euglena gracilis* bei Mixotrophie und Autotrophie. *Planta* **70**, 344–389.

Britz, S. J., and Briggs, W. R. (1976). Circadian rhythms of chloroplast orientation and photosynthetic capacity in *Ulva. Plant Physiol.* **58**, 22–27.

Britz, S. J., and Briggs, W. R. (1983). Rhythmic chloroplast migration in the green alga *Ulva*: dissection of movement mechanism by differential inhibitor effects. *Eur. J. Cell Biol.* **31**, 1–8.

Broda, H., and Schweiger H.-G. (1981). Long-term measurement of endogenous diurnal oscillations of the electrical potential in an individual *Acetabularia* cell. *Eur. J. Cell Biol.* **26**, 1–4.

Broda, H., Schweiger, G., Koop, H.-U., Schmid, R., and Schweiger, H.-G. (1979). Chloroplast migration: A method for continuously monitoring a circadian rhythm in a single cell of *Acetabularia. In* "Developmental Biology of *Acetabularia*" (S. Bonoto, V. Kefelt, and S. Puiseux-Dao, eds.), pp. 163–167. Elsevier/North Holland Biomedical Press, Amsterdam.

Broda, H., Brugge, D., Homma, K., and Hastings, J. W. (1986). Circadian communication between unicells? Effects on period by cell-conditioning of medium. *Cell Biophys.* **8**, 47–67.

Bruce, V. G. (1970). The biological clock in *Chlamydomonas reinhardi. J. Protozool.* **17**, 328–334.

Bruce, V. G. (1972). Mutants of the biological clock in *Chlamydomonas reinhardi. Genetics* **70**, 537–548.

Bruce, V. G., and Pittendrigh, C. S. (1956). Temperature independence in a unicellular "clock." *Proc. Natl. Acad. Sci. U. S. A.* **42**, 676–682.

Brun, N. A. (1962). Rhythmic stomatal opening responses in banana leaves. *Physiol. Plant.* **15**, 623–630.

Bryant, T. R. (1973). 24-hour periodicity of gas analysis in dry seeds: Circadian rhythmicity in the absence of respiration, transpiration and translation. *Science* **178**, 634.

Buchanan-Bollig, I. C., Fischer, A., and Kluge, M. (1984). Circadian rhythms in *Kalanchoe:* The pathway of $^{14}CO_2$ fixation during prolonged light. *Planta* **161**, 71–80.

Bühnemann, F. (1955a). Die rhythmische Sporebildung von *Oedogonium cardiacum* Wittr. *Biol. Zentralbl.* **74**, 1–54.

Bühnemann, F. (1955b). Das endodiurnale System der *Oedogonium*-Zelle. III. Uber den Temperatureinfluss. *Z. Naturforsch., B: Anorg. Chem., Org. Chem., Biochem., Biophys., Biol.* **10B**, 305–310.

Bünning, E. (1931). Untersuchungen uber die autonomen tagesperiodischen Bewegungen der Primarblatter von *Phaseolus multiflorus. Jahrb. Wiss. Bot.* **25**, 439–480.

Bünning, E., and Tazawa, M. (1957). Uber den Temperatureinfluss auf die endogene Tagesrhythmik bei *Phaseolus. Planta* **50**, 107–121.

Bünsow, R. C. (1960). The circadian rhythm of photoperiodic responsiveness in *Kalanchoe. Cold Spring Harbor Symp. Quant. Biol.* **25**, 257–260.

Bush, K. J., and Sweeney, B. M. (1972). The activity of ribulose diphosphate carboxylase in extracts of *Gonyaulax polyedra* in the day and the night phase of the circadian rhythm of photosynthesis. *Plant Physiol.* **50**, 446–451.

Chen, Y. B., Lee, Y., and Satter, R. L. (1984). Chronobiology of aging in *Albizzia julibrissin.* I. An automated computerized system for monitoring leaflet movement, the rhythm in constant darkness. *Plant Physiol.* **76**, 858–860.

Chisholm, S. W., and Brand, L. E. (1981). Persistence of cell division phasing in marine phytoplankton in continuous light after entrainment in light : dark cycles. *J. Exp. Mar. Biol.* **51**, 107–118.

Dunlap, J. C., and Hastings, J. W. (1981). The biological clock in *Gonyaulax* controls luciferase activity by regulating turnover. *J. Biol. Chem.* **256**, 10509–10518.

Edmunds, L. N., Jr. (1966). Studies on synchronously dividing cultures of *Euglena gracilis* Klebs (strain Z). *J. Cell. Physiol.* **67**, 35–44.

Edmunds, L. N., Jr., and Funch, R. (1969a). Effects of "skeleton" photoperiods and high frequency light-dark cycles on the rhythm of cell division in synchronized cultures of *Euglena. Planta* **87**, 134–163.

Edmunds, L. N., Jr., and Funch, R. R. (1969b). Circadian rhythm of cell division in *Euglena*: Effects of a random illumination cycle. *Science* **165**, 500–503.

Edmunds, L. N., Jr., Tay, D. E., and Laval-Martin, D. L. (1982). Cell division cycles and circadian clocks. Phase-response curves for light perturbations in synchronous cultures of *Euglena. Plant Physiol.* **70**, 297–302.

Engelmann, W. (1960). Endogene Rhythmik und photoperiodische Bluhinduktion bei *Kalanchoe. Planta* **55**, 496–511.

Engelmann, W., Eger, I., Johnsson, A., and Karlsson, H. G. (1974). Effect of temperature pulses on the petal rhythm of *Kalanchoe*: An experimental and theoretical study. *Int. J. Chronobiol.* **2**, 347–358.

Engelmann, W., Johnsson, A., Karlsson, H. G., Kobler, R., and Schimmel, M. L. (1978). Attentuation of the petal movement rhythm in *Kalanchoe* with light pulses. *Physiol. Plant.* **43**, 68–76.

Fitt, M. K., Chang, S. S., and Trench, R. K. (1981). Motility patterns of different strains of the symbiotic dinoflagellate, *Symbiodinium microadriaticum* (Freudenthal) in culture. *Bull. Mar. Sci.* **31**, 436–443.

Forward, R. B., Jr., and Davenport, D. (1970). The circadian rhythm of a behavioral photoresponse in the dinoflagellate *Gyrodinium dorsum*. *Planta* **92**, 259–266.

Gordon, W. L., and Koukkari, W. L. (1978). Circadian rhythmicity in the activities of phenylalanine ammonia lyase from *Lemna perpusilla* and *Spirodela polyrhiza*. *Plant Physiol.* **62**, 612–615.

Gorton, H. L., and Satter, R. L. (1984). Extensor and flexor protoplasts from *Samanea* pulvini. I. Isolation and initial characterization. *Plant Physiol.* **76**, 680–684.

Goto, K. (1984). Causal relationships among metabolic circadian rhythms in *Lemna*. *Z. Naturforsch., C: Biosci.* **39C**, 73–84.

Grobbelaar, N., Huang, T. C., Lin, H. Y., and Chow, T. J. (1986). Dinitrogen-fixing endogenous rhythm in *Synechococcus* RF-1. *FEMS Microbiol. Let.* **37**, 173–177.

Grossenbacher, K. A. (1939). Autonomic cycle of rate of exudation in plants. *Am. J. Bot.* **26**, 107–109.

Halberg, F., Hastings, J. W., Cornelisson, G., and Broda, H. (1985). *Gonyaulax polyedra* "talks" both "circadian" and "circaseptan." *Chronobiologia (Milan)* **12**, 185.

Hartwig, R., Schweiger, M., Schweiger, R., and Schweiger, H.-G. (1985). Identification of a high molecular weight polypeptide that may be part of the circadian clockwork in *Acetabularia*. *Proc. Natl. Acad. Sci. U. S. A.* **82**, 163–167.

Hastings, J. W., and Bode, V. C. (1962). Biochemistry of rhythmic systems. *Ann. N. Y. Acad. Sci.* **98**, 876–889.

Hastings, J. W., and Sweeney, B. M. (1957). On the mechanism of temperature independence in a biological clock. *Proc. Natl. Acad. Sci. U. S. A.* **43**, 804–811.

Hastings, J. W., and Sweeney, B. M. (1958). A persistent diurnal rhythm of luminescence in *Gonyaulax polyedra*. *Biol. Bull. (Woods Hole, Mass.)* **115**, 440–458.

Hastings, J. W., and Sweeney, B. M. (1960). The action spectrum for shifting the phase of the rhythm of luminescence in *Gonyaulax polyedra*. *J. Gen. Physiol.* **43**, 697–706.

Hastings, J. W., and Sweeney, B. M. (1964). Phase cell division in the marine dinoflagellates. *In* "Synchrony in Cell Division and Growth" (E. Zeuthen, ed.), pp. 307–321. Wiley, New York.

Hastings, J. W., Astrachan, L., and Sweeney, B. M. (1961). A persistent daily rhythm in photosynthesis. *J. Gen. Physiol.* **45**, 69–76.

Haxo, F. T., and Sweeney, B. M. (1955). Bioluminescence in *Gonyaulax polyedra*. *In* "The Luminescence of Biological Systems" (F. H. Johnson, ed.), pp. 415–420. Am. Assoc. Adv. Sci., Washington, D. C..

Herman, E. M., and Sweeney, B. M. (1975). Circadian rhythm of chloroplast ultrastructure in *Gonyaulax polyedra*, concentric organization around a central cluster of ribosomes. *J. Ultrastruct. Res.* **50**, 347–354.

Herman, E. M., and Sweeney, B. M. (1976). *Cachonina illdefina* sp. nov. (Dinophyceae): Chloroplast tubules and degeneration of the pyrenoid. *J. Phycol.* **12**, 198–205.

Hesse, M. (1972). Endogenous rhythm of the productivity in *Chlorella* and the influence of light. *Z. Pflanzenphysiol.* **67**, 58–77.

Hew, C. S., Thio, Y. C., Wong, S. Y., and Chen, T. Y. (1978). Rhythmic production of CO_2 by tropical orchid flowers. *Physiol. Plant.* **42**, 226–230.

Hillman, W. S. (1964). Endogenous circadian rhythms and the response of *Lemna perpusilla* to skeleton photoperiods. *Am. Nat.* **98**, 323–328.

Hohman, T. G. (1972). Diurnal periodicity in the photosynthetic activity of *Caulerpa racemosa* (Forsskal) J. Ag. *J. Phycol.* **8**, 16.

Hoshizake, T., and Hamner, K. C. (1964). Circadian leaf movements: Persistence in bean plants grown in high-intensity light. *Science* **144**, 1240–1241.

Ibrahim, C. A., Lecharny, A., and Millet, B. (1981). Circadian endogenous growth rhythm in tomato. *Plant Physiol.* **67**, 113.

Iglesias, A., and Satter, R. L. (1983). H^+ fluxes in excised *Samanea* motor tissue. II. Rhythmic properties. *Plant Physiol.* **72**, 570–572.

Jerebzoff, S., and Vanden Driessche, T. (1983). Rythme circadien de la teneur en acides gras dans la cellule entière et les chloroplastes d'*Acetabularia mediterranea*. *C. R. Seances Acad. Sci.* **296**, 319–322.

Johnson, C. H., Roeber, J. F., and Hastings, J. W. (1984). Circadian changes in enzyme concentration account for rhythm of enzyme activity in *Gonyaulax*. *Science* **223**, 1428–1430.

Kageyama, A., Yokohama, Y., and Nisizawa, K. (1979). Diurnal rhythm of apparent photosynthesis of a brown alga, *Spatoglossum pacificum*. *Bot. Mar.* **22**, 199–201.

Karakashian, M. W., and Schweiger, H. G. (1976). Circadian properties of the rhythmic system in individual nucleated and enucleated cells of *Acetabularia meditteranea*. *Exp. Cell Res.* **97**, 366–377.

Klemm, E., and Ninnemann, H. (1976). Detailed action spectrum for the delay phase shift in pupal emergence of *Drosophila pseudoobscura*. *Photochem. Photobiol.* **24**, 369–372.

Kloppstech, K. (1985). Diurnal and circadian rhythmicity in the expression of light-induced plant nuclear RNAs. *Planta* **165**, 502–506.

Kondo, T., and Tsudzuki, T. (1978). Rhythm in potassium uptake by a duckweed, *Lemna gibba* G3. *Plant Cell Physiol.* **19**, 1465–1473.

Koukkari, W. L., Duke, S. H., Halberg, F., and Lee, J.-K. (1974). Circadian rhythmic leaflet movements: Student exercise in chronobiology. *Chronobiologia (Milan)* **1**, 281–302.

Laval-Martin, D. L., Shuch, D. J., and Edmunds, L. N., Jr. (1979). Cell cycle-related and endogenously controlled circadian photosynthetic rhythms in *Euglena. Plant Physiol.* **63**, 495–502.

Lecharny, A., and Wagner, E. (1984). Stem extension rate in light-grown plants. Evidence for an endogenous circadian rhythm in *Chenopodium rubrum. Physiol. Plant.* **60**, 437–443.

Leinweber, F. J. (1961). Temperature coefficient of endodiurnal leaf movements in *Phaseolus. Nature (London)* **189**, 1028.

Leong, T.-Y., and Schweiger, H.-G. (1979). The role of chloroplast-membrane protein synthesis in the circadian clock. Purification and partial characterization of a polypeptide which is suggested to be involved in the clock. *Eur. J. Biochem.* **98**, 187–194.

Lerch, K. E., and Cook, C. B. (1984). Some effects of photoperiod on the motility rhythm of cultured zooxanthellae. *Bull. Mar. Sci.* **34**, 477–483.

Lonergan, T. A. (1981). A circadian rhythm in the rate of light-induced electron flow in three leguminous species. *Plant Physiol.* **68**, 1041–1046.

Lonergan, T. A., and Sargent, M. L. (1979). Regulation of the photosynthesis rhythm in *Euglena gracilis.* 2. Involvement of electron flow through both photosystems. *Plant Physiol.* **64**, 99–103.

Macdowall, F. D. H. (1964). Reversible effects of chemical treatments on the rhythmic exudation of sap by tobacco roots. *Can. J. Bot.* **42**, 115–122.

Malinowski, J. R., Laval-Martin, D. L., and Edmunds, L. N., Jr. (1985). Circadian oscillators, cell cycles, and singularities: Light perturbations of the free-running rhythm of cell division in *Euglena. J. Comp. Physiol.* **155**, 257–267.

Meeson, B. W. (1977). Circadian rhythmicity in the marine dinoflagellate, *Ceratium furca. J. Phycol.* **13s**, 45.

Mergenhagen, D., and Schweiger, H. G. (1975). Circadian rhythm of oxygen evolution in cell fragments of *Acetabularia mediterranea. Exp. Cell Res.* **92**, 127–130.

Minder, C., and Vanden Driessche, T. (1978). Changes in the cyclic AMP content during growth and development of *Acetabularia. Differentiation (Berlin)* **10**, 165–170.

Miyata, H., and Yamamoto, Y. (1969). Rhythms in respiratory metabolism of *Lemna gibba* G3 under continuous illumination. *Plant Cell Physiol.* **10**, 875–889.

Nakashima, H. (1976). Diurnal rhythm of uridine incorporation into RNA regulated by two light-perceiving systems in a long-day duckweed, *Lemna gibba* G3. *Plant Cell Physiol.* **17**, 209–217.

Nakashima, H. (1979). Diurnal rhythm of nuclear RNA polymerase I activity in a duckweed, *Lemna gibba* G3 under continuous light conditions. *Plant Cell Physiol.* **20**, 165–176.

Nimmo, G. A., Nimmo, H. G., Fewson, C. A., and Wilkins, M. B. (1984).

Diurnal changes in the properties of phosphoenolpyruvate carboxylase in *Bryophyllum* leaves: A possible covalent modification. *FEBS Lett.* **178**, 199–203.

Novak, B., and Greppin, H. (1979). High-frequency oscillations and circadian rhythm of the membrane potential in spinach leaves. *Planta* **144**, 235–240.

Novak, B., and Sironval, C. (1976). Circadian rhythm of transcellular current in regenerating enucleate posterior stalk segments of *Acetabularia mediterranea. Plant Sci. Lett.* **6**, 273–284.

Nultsch, W., Ruffer, U., and Pfau, J. (1984). Circadian rhythms in the chromatophore movements of *Dictyota dichotoma. Mar. Biol. (Berlin)* **81**, 217–222.

Okada, M., Inoue, M., and Ikeda, T. (1978). Circadian rhythm in photosynthesis of the green alga *Bryopsis maxima. Plant Cell Physiol.* **19**, 197–202.

O'Leary, M. H. (1982). Phosphoenolpyruvate carboxylase: An enzymologist's view. *Annu. Rev. Plant Physiol.* **33**, 297–315.

Oohusa, T. (1980). Diurnal rhythm in the rates of cell division, growth and photosynthesis of *Porphyra yezoensi* (Rhodophyceae) cultured in the laboratory. *Bot. Mar.* **23**, 1–5.

Östgaard, K., and Jensen, A. (1982). Diurnal and circadian rhythms in the turbidity of growing *Skeletonema costatum* cultures. *Mar. Biol. (Berlin)* **66**, 261–268.

Overland, L. (1960). Endogenous rhythm in opening and odor of flowers of *Cestrum nocturnum. Am. J. Bot.* **47**, 378–382.

Pohl, R. (1948). Tagesrhythmus im phototaktischen Verhalten der *Euglena gracilis. Z. Naturforsch., B: Anorg. Chem., Org. Chem., Biochem., Biophys., Biol.* **3B**, 367–378.

Prézelin, B. B., Meeson, B. W., and Sweeney, B. M. (1977a). Characterization of photosynthetic rhythms in marine dinoflagellates. I. Pigmentation, photosynthetic capacity and respiration. *Plant Physiol.* **60**, 384–387.

Prézelin, B. B., Meeson, B. W., and Sweeney, B. M. (1977b). Characterization of photosynthetic rhythms in marine dinoflagellates. II. Photosynthesis-irradiance curves and *in vivo* chlorophyll a fluorescence. *Plant Physiol.* **60**, 388–392.

Queiroz, O. (1974). Circadian rhythms and metabolic patterns. *Annu. Rev. Plant Physiol.* **25**, 115–134.

Rensing, L., Taylor, W., Dunlap, J., and Hastings, J. W., (1980). The effect of protein synthesis inhibitors on the *Gonyaulax* clock. II. The effect of cycloheximide on ultrastructural parameters. *J. Comp. Physiol.* **138**, 9–18.

Richter, G. (1963). Die Tagesperiodik der Photosynthese bei *Acetabularia* und ihre Abhangigkeit von Kernaktivitat, RNS- und Protein-Synthese. *Z. Naturforsch., B: Anorg. Chem., Org. Chem., Biochem., Biophys., Biol.* **18B**, 1085–1089.

Ruddat, M. (1961). Versuche zur Beeinflussung und Auslosung der endogenen Tagesrhythmik bei *Oedogonium cardiacum* Wittr. *Z. Bot.* **49**, 23–46.

Samuelsson, G., Sweeney, B. M., Matlick, H. A., and Prezelin, B. B. (1983). Changes in photosystem II account for the circadian rhythm in photosynthesis in *Gonyaulax polyedra. Plant Physiol.* **73**, 329–331.

Satter, R. L. (1979). Leaflet movements and tendril curling. *Encycl. Plant Physiol., New Ser.* **7**, 442–484.

Satter, R. L., Geballe, G. T., Applewhite, P. B., and Galston, A. W. (1974). Potassium flux and leaf movement in *Samanea saman.* I. Rhythmic movement. *J. Gen. Physiol.* **64**, 413–430.

Satter, R. L., Guggino, S. E., Lonergan, T. A., and Galston, A. W. (1981). The effects of blue and far red light on rhythmic leaflet movements in *Samanea* and *Albizzia. Plant Physiol.* **67**, 965–968.

Schrempf, M., and Mayer, W.-E. (1980). Electron microprobe analysis of the circadian changes in K and Cl distribution in the laminar pulvinus of *Phaseolus coccineus* L. *Z. Pflanzenphysiol.* **100**, 247–253.

Schweiger, E., Wallraff, H. G., and Schweiger, H. G. (1964). Uber tagesperiodische Schwankungen der Sauerstoffebilanz kernhaltiger und kernloser *Acetabularia mediterranea. Z. Naturforsch., B: Anorg. Chem., Org. Chem., Biochem., Biophys., Biol.* **19B**, 499–505.

Scott, B. I. H., and Gulline, H. F. (1972). Natural and forced circadian oscillations in the leaf of *Trifolium repens. Aust. J. Biol. Sci.* **25**, 61–76.

Simon, E., Satter, R. L., Galston, A. W. (1976a). Circadian rhythmicity in excised *Samanea* pulvini. I. Sucrose-white light interactions. *Plant Physiol.* **58**, 417–420.

Simon, E., Satter, R. L., and Galston, A. W. (1976b). Circadian rhythmicity in excised *Samanea* pulvini. II. Resetting the clock by phytochrome conversion. *Plant Physiol.* **59**, 421–425.

Spudich, J. L., and Sager, R. (1980). Regulation of the *Chlamydomonas* cell cycle by light and dark. *J. Cell Biol.* **85**, 136–145.

Stålfelt, M. G. (1965). The relation between the endogenous and induced elements of the stomatal movements. *Physiol. Plant.* **18**, 177–184.

Straley, S. C., and Bruce, V. G. (1979). Stickiness to glass. Circadian changes in the cell surface of *Chlamydomonas reinhardi. Plant Physiol.* **63**, 1175–1181.

Sweeney, B. M. (1960). The photosynthetic rhythm in single cells of *Gonyaulax polyedra. Cold Spring Harbor Symp. Quant Biol.* **25**, 145–148.

Sweeney, B. M. (1976). A freeze-fracture study of *Gonyaulax polyedra.* I. Membranes associated with the theca and circadian changes in the particles of one membrane face. *J. Cell Biol.* **68**, 451–461.

Sweeney, B. M. (1979). Bright light does not immediately stop the circadian clock of *Gonyaulax. Plant Physiol.* **64**, 341–344.

Sweeney, B. M. (1981a). Freeze-fractured chloroplast membranes in *Gonyaulax polyedra* (Pyrrophyta). *J. Phycol.* **17**, 95–101.

Sweeney, B. M. (1981b). The circadian rhythms in bioluminescence, photosynthesis and organellar movement in the large dinoflagellate, *Pyrocystis fusiformis. In* "International Cell Biology 1980-1981" (H.-G. Schweiger, ed.), pp. 807–814. Springer-Verlag, Berlin and New York.

Sweeney, B. M. (1982). Interaction of the circadian cycle with the cell cycle in *Pyrocystis fusiformis. Plant Physiol.* **70**, 272–276.

Sweeney, B. M. (1983). Circadian time-keeping in eukaryotic cells, models and experiments. *Prog. Phycol. Res.* **2**, 189–225.

Sweeney, B. M., and Hastings, J. W. (1957). Characteristics of the diurnal rhythm of luminescence in *Gonyaulax polyedra. J. Cell. Comp. Physiol.* **49**, 115–128.

Sweeney, B. M., and Hastings, J. W. (1958). Rhythmic cell division in populations of *Gonyaulax polyedra. J. Protozool.* **5**, 217–224.

Sweeney, B. M., and Haxo, F. T. (1961). Persistence of a photosynthetic rhythm in enucleated *Acetabularia. Science* **134**, 1361–1363.

Swift, E., and Taylor, W. R. (1967). Bioluminescence and chloroplast movement in the dinoflagellate *Pyrocystis lunula. J. Phycol.* **3**, 77–81.

Takahashi, J. S., DeCoursey, P. J., Bauman, L., and Menaker, M. (1984). Spectral sensitivity of a novel photoreceptive system mediating entrainment of mammalian circadian rhythms. *Nature (London)* **308**, 186–188.

Terborgh, J., and McLeod, G. C. (1967). The photosynthetic rhythm of *Acetabularia crenulata.* I. Continuous measurements of oxygen exchange in alternating light-dark regimes and in constant light of different intensities. *Biol. Bull. (Woods Hole, Mass.)* **133**, 659–669.

Terry, O. W., and Edmunds, L. N., Jr. (1970). Phasing of cell division by temperature cycles in *Euglena* cultured autotrophically under continuous illumination. *Planta* **93**, 106–127.

Topperwein, F. J., and Hardeland, R. (1980). Free-running circadian rhythm of plastid movements in individual cells of *Pyrocystis lunula* (Dinophyta). *J. Interdiscip. Cycle Res.* **11**, 325–329.

Uhlig, G., and Sahling, G. (1982). Rhythms and distributional phenomena in *Noctiluca miliaris. Ann. Inst. Oceanogr. (Paris)* **58**, 277–284.

Vanden Driessche, T. (1966a). Circadian rhythms in *Acetabularia*: Photosynthetic capacity and chloroplast shape. *Exp. Cell Res.* **42,** 18–30.

Vanden Driessche, T. (1966b). The role of the nucleus in the circadian rhythm of *Acetabularia mediterranea. Biochim. Biophys. Acta* **126,** 456–470.

Vanden Driessche, T. (1970). Circadian variation in ATP content in the chloroplasts of *Acetabularia mediterranea. Biochim. Biophys. Acta* **205**, 526–528.

Vanden Driessche, T., and Bonotto, S. (1969). The circadian rhythm in RNA synthesis in *Acetabularia mediterranea. Biochim. Biophys. Acta* **179**, 58–66.

Vanden Driessche, T., and Hars, R. (1972a). Variations circadiennes de l'ultrastructure des chloroplastes d'*Acetabularia.* 1. Algues entières. *J. Microsc. (Paris)* **15**, 85–90.
Vanden Driessche, T., and Hars, R. (1972b). Variations circadiennes de l'ultrastructure des chloroplastes l'*Acetabularia.* II. Algues anucléées. *J. Microsc. (Paris)* **15**, 91–98.
Waaland, S. D., and Cleland, R. (1972). Development in the red alga, *Griffithsia pacifica.* Control by internal and external factors. *Planta* **105**, 196–204.
Wagner, E., Stroebele, L., and Frosch, S. (1974). Endogenous rhythmicity and energy transduction. V. Rhythmicity in adenine nucleotides and energy charge in seedlings of *Chenopodium rubrum. J. Interdiscip. Cycle Res.* **5**, 77–88.
Walz, B., Walz, A., and Sweeney, B. M. (1983). A circadian rhythm in RNA in the dinoflagellate, *Gonyaulax polyedra. J. Comp. Physiol. B* **151B**, 207–213.
Warren, D. M., and Wilkins, M. B. (1961). An endogenous rhythm in the rate of dark fixation of carbon dioxide in leaves of *Bryophyllum fedtschenkoi. Nature (London)* **191**, 686–688.
Wilkerson, M. J., and Smith, H. (1976). Properties of phosphoenol pyruvate carboxylase from *Bryophyllum fedtschenkoi* leaves and fluctuations in carboxylase activity during the endogenous rhythm of carbon dioxide output. *Plant Sci. Lett.* **6**, 319–324.
Wilkins, M. B. (1959). An endogenous rhythm in the rate of carbon dioxide output of *Bryophyllum.* I. Some preliminary experiments. *J. Exp. Bot.* **10**, 377–390.
Wilkins, M. B. (1960). An endogenous rhythm in the rate of carbon dioxide output of *Bryophyllum.* II. The effect of light and darkness on the phase and period of the rhythm. *J. Exp. Bot.* **11**, 269–288.
Wilkins, M. B. (1962). An endogenous rhythm in carbon dioxide output of *Bryophyllum.* III. The effects of temperature on the phase and the period of the rhythm. *Proc. R. Soc. London, Ser. B* **156**, 220–241.
Wilkins, M. B. (1973). An endogenous circadian rhythm in the rate of carbon dioxide output in *Bryophyllum.* VI. Action spectrum for the induction of phase shifts by visible radiation. *J. Exp. Bot.* **24**, 488–496.
Wilkins, M. B. (1983). The circadian rhythm of carbon-dioxide metabolism in *Bryophyllum*: The mechanism of phase-shift induction by thermal stimuli. *Planta* **157**, 471–480.
Wilkins, M. B., and Holowinsky, A. W. (1965). The occurrence of an endogenous circadian rhythm in a plant tissue culture. *Plant Physiol.* **40**, 907–909.
Zimmer, R. (1962). Phasenverschienbung und andere Storlichtwirkungen auf die endogen tagesperiodischen Blutenblattbewegungen von *Kalanchoe blossfeldiana. Planta* **58**, 283–300.

Mechanism of Circadian Timing

The phenomenon of rhythmicity provides clear evidence that both plants and animals can tell time. But how do they do it? This has proven to be a difficult question to answer. Since we know much more about the circadian rhythms than about the rhythms with either shorter or longer periods, most of the research aimed at understanding the mechanism of the "clock" has considered circadian timing and the experiments to answer questions about mechanism have used circadian systems. In this chapter I shall try to summarize what has been accomplished and point out what is still to be done before we can understand the mechanism of the circadian clock in any cells.

For some time the idea persisted that there might be some unidentified environmental cycle that provided time information to plants and animals. The rotation of the earth with a 24-h period was thought to be the cause of this cycle in some unknown way, since light and temperature fluctuations could be held constant in the laboratory and circadian rhythms still persisted, thus eliminating these environmental cycles as timers (Brown *et al.*, 1970). This idea has now been laid to rest by experiments that demonstrated circadian rhythms both at the south pole

(Hamner *et al.*, 1962) and in space (Sulzman *et al.*, 1984). The "clock" must then be within the cells, not outside. But where?

One word of caution — it is dangerous to apply the experimental results from one organism to another. There is no assurance at present that the clock in all cells is the same, although this is an intuitive generalization. Let us be cautious not to fall into this potential trap. I shall first discuss what is known about the mechanism of time-keeping in the dinoflagellates and then consider whether can generalize from these to other organisms.

ONE CLOCK PER CELL?

A single isolated *Gonyaulax* clearly has a clock. Rhythmicity in both bioluminescence (Krasnow *et al.*, 1981) and photosynthesis (Sweeney, 1960) can be seen in a single cell. The question is then, "Does a *Gonyaulax* cell have a single clock or more than one?" *Gonyaulax* has the advantage that it shows a number of circadian rhythms, called "overt" because they can be observed and measured, for example, the circadian rhythms in bioluminescence and photosynthesis. Neither is the time-keeping oscillation, since alterations in the rates of these processes have no effect on timing: there is no feedback from the overt rhythm to the clock. The presence of several overt rhythms in one organism makes it possible to test whether each cell has a single clock or more than one. If overt rhythms all keep the same period during long times under constant conditions and respond with the same phase-shift to a given stimulus, a light pulse, for example, then the simplest explanation is that all these rhythms are controlled by a single clock. When the rhythms in bioluminescence, photosynthesis, and cell division were measured in the same cell population over 12 days in LL, it was clear that all the rhythms were in exactly the same phase relationship at the end as at the beginning of this time (McMurray and Hastings, 1972). Furthermore, when short pulses of ultraviolet light were given to a culture (Sweeney, 1963), both the rhythms of bioluminescence and cell division were phase-shifted by the same amount (Fig. 4.1). Until

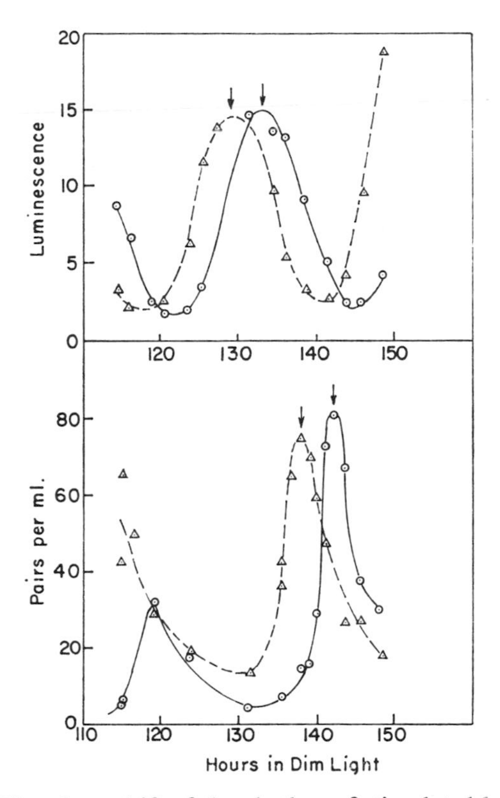

Fig. 4.1. The phase shift of the rhythm of stimulated luminescence in *Gonyaulax polyedra* in response to ultraviolet light (2 min, 150 ergs cm^{-2} sec^{-1} at time 0). The upper curves show the rhythm in stimulated luminescence; the lower curves, the rhythm in cell division; dashed curves, irradiated cells; solid curves, unirradiated controls (Sweeney, 1963).

evidence to the contrary is found, a single clock per cell is the simplest assumption.

INHIBITORS AND PHASE SHIFTS

One clock, then, but what are its components? How can we recognize them or distinguish them from the overt rhythmic processes? Since overt rhythms by definition do not feed back to the clock, we can look for processes that do show a feedback, i.e., a change in timing when rates are either increased or decreased. It can be argued that such processes must be close to or a part of the clock mechanism.

When we began to plan such experiments, the only properties of the clock that we could detect were its period and phase, since we knew nothing of its biochemistry. The plan was then to treat cells in constant light conditions with a substance that affects a known biochemical path and measure the period or the phase of an overt circadian rhythm compared with that of an untreated control cell population. Many experiments of this type have been done (Hastings and Schweiger, 1976). There is of course the uncertainty as to whether or not the substance used really has the specific effect that we assign to it. "Only the uninhibited use inhibitors," but this technique provided at least a starting point. Most of these experiments were designed to detect changes in the phase of a circadian rhythm. Usually in our work this meant looking for changes in the phase of the circadian rhythm in bioluminescence, because the measurement of light emission is both quick and simple. Since pulses of light of short duration relative to 24 h were known to shift the phase of many circadian rhythms in a characteristic way, perhaps inhibitors might do the same. Exposing cells to inhibitors for long times to detect changes in period often causes toxicity. Furthermore, the period is difficult to measure unless you have a device to record automatically, since a long sequence of measurements are required for an accurate determination of the period.

What have we learned from such experiments? The first results were quite discouraging since various inhibitors of metabolism had no consistent effect on the phase of the rhythm in bioluminescence in *Gonyaulax* (Hastings, 1960). The inhibitors used were azide, cyanide, dinitrophenol, *para*-chloromercuribenzoate, urethane, and arsenite. The 8-h treatments were staggered to cover 24 h. Any simple mechanism invoking energy-generating intermediary metabolism was eliminated by the failure of these metabolic poisons to affect circadian timing.

When actinomycin D, a compound that binds to DNA, became available, it was assayed for effects on the phase and period of the rhythm in bioluminescence in *Gonyaulax* (Karakashian and Hastings, 1962, 1963). This substance proved to be toxic and hence difficult to work with. Actinomycin markedly inhibited the rise in bioluminescence. Furthermore, short exposures were not possible, since the drug could not be washed out, once added to a cell suspension. After some days, the bioluminescence rhythm reappeared, albeit at very low amplitude. The phase and period was unchanged compared to untreated control. When the rhythm in photosynthesis was assayed with actinomycin, 10 times higher concentrations were required than with bioluminescence to cause any inhibition. These findings did not favor a clock that required transcription.

Ionophores, which alter the permeability of membranes to ions, were the next substances to be assayed. Pulses of the K^+ ionophore valinomycin caused rather small but distinct phase shifts of the rhythm in bioluminescence in *Gonyaulax*, and, perhaps more significant, the amount of phase shift depended on the time in the cycle when the substances were administered, positive phase shifts being greatest toward the end of subjective day and negative phase shifts occurring near midnight (Sweeney, 1974a). Notice that while the shape of the phase response curve (PRC) resembled that for light pulses, the whole curve was displaced in time by about 12 h (Fig. 4.2), thus resembling the PRC in response to dark pulses in a light environment (Karakashian and Schweiger, 1976). As would be expected were potassium ions important in time-keeping, the K^+ concentration in the cells showed a circadian rhythmicity, highest at CT 06 and lowest at CT 18 (Sweeney, 1974a). Experiments with gramicidin, an Na^+ and K^+ ionophore, and A23187, a Ca^{2+} ionophore, gave negative results (Sweeney, 1976b; Sweeney and Herz, 1977).

The results of the experiments with valinomycin generated two models of the circadian clock. Both emphasized the importance of membranes in the clock mechanism. They both suggested a two-component feedback loop, composed of a potassium gradient across a membrane and transport of potassium across a membrane. In one model this membrane was postulated

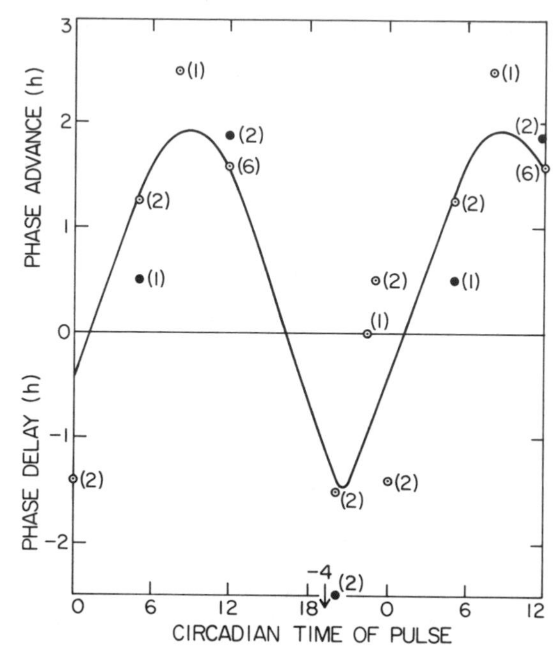

Fig. 4.2. Phase-response curve for the effect of 4-h exposure to valinomycin, 2×10^{-7} *M*, dissolved in acetone, final concentration 0.1%, on the rhythm in stimulated bioluminescence in *Gonyaulax polyedra*. The value of *n* is given in parentheses beside each point. Open circles, valinomycin without extra K^+; closed circles, valinomycin + 50 m*M* extra K^+ as KCl. LL, 580 lux, 20–21°C. Double plotted for clarity (Sweeney and Herz, 1977).

to be the plasma membrane (Njus *et al.,* 1974), while in the other, organelle membranes played this role (Sweeney, 1974b). The ion gradient affected the membrane transport to complete the feedback loop. These models accounted nicely for the phase-shifting effects of light if one assumed that light opened the K^+ channels in the membrane, setting the ion gradient to zero. It is not clear how a gradient between the cytoplasm and the essentially infinite cell environment could cycle. The manner in which the state of the membrane changed was not clear either, since lateral diffusion of membrane transport molecules in the membrane, suggested by Njus *et al.* (1974) is too fast to play this role. In fact, explaining the long period has been a difficulty for all models so far. Resets were also obtained in experiments with pulses of ethanol (Sweeney, 1974a; Taylor *et al.,* 1979), and procaine and dibucaine (Walz, 1981), substances that may well affect membranes, although their mode of action is not clear. That a membrane effect of alcohols was responsible for the phase shifts caused by these substances seemed unlikely because short-chain alcohols were more effective in resetting than long-chain alcohols (Sweeney and Herz, 1977), while the reverse is usually found when alcohols are known to be affecting membranes. A possible metabolic product of ethanol, acetaldehyde, has been found to shift phase in *Gonyaulax* (Taylor and Hastings, 1979), but alcohols do not serve as a carbon source for this organism. An interesting mathematical model has been studied by Chay (1981), in which the ion is H^+ and transport in and out of the mitochondria is postulated, combined with the pH dependency of mitochondrial enzymes.

The membrane potential in *Gonyaulax* measured with the cyanine dye diS-3C(5), which changes its fluorescence as a function of membrane potential, was shown to oscillate with a circadian rhythm (Adamich *et al.,* 1976). It was not possible to determine which membrane gave rise to the potential being measured in these experiments. Vanillic acid, which was shown to alter membrane potential in *Gonyaulax* (Kiessig *et al.,* 1979), shifted phase, as would be expected from a membrane model, since the membrane potential depends on an ion gradient. One component of the feedback loop is thus likely to involve an ion, probably K^+. But what are the other components?

Certain membrane-bound proteins are known to be transport molecules controlling the migration of ions across membranes. Experiments testing the effect on *Gonyaulax* of inhibitors of protein synthesis led to the interesting result that, while pulses of chloramphenicol, which inhibits translation on 70S ribosomes, did not cause phase shifts, pulses with the 80S translation inhibitors cycloheximide and anisomycin (Figs. 4.3 and 4.4), caused large and fully reproducible resetting of the circadian rhythm in bioluminescence (Walz and Sweeney, 1979; Dunlap *et al.,* 1980;

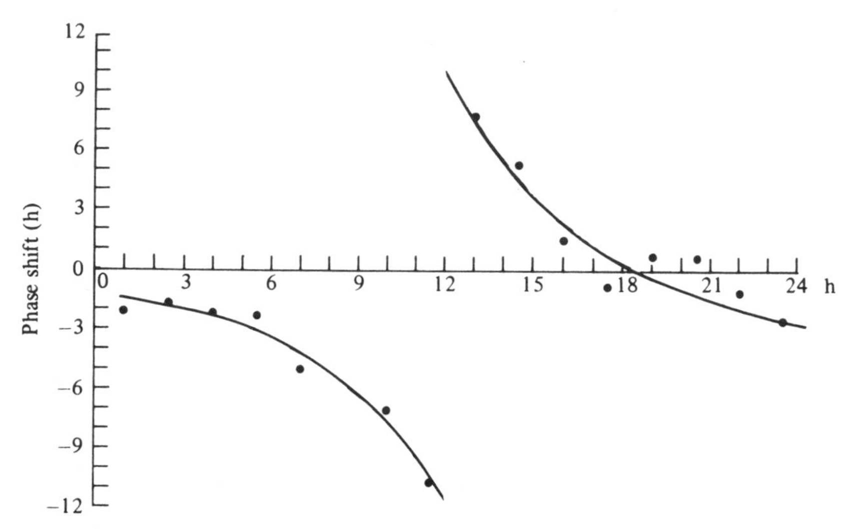

Fig. 4.3. Phase-response curve for *Gonyaulax polyedra* glow rhythm treated with 300 n*M* anisomycin for 1 h at different times in a circadian cycle. Points are plotted at the beginning of the inhibitor pulse, not the middle of the pulse as in other PRCs illustrated here. Temperature, 19°C, light, 35 μE m^{-2}, s^{-1} (Taylor, *et al.*, 1982b).

Taylor *et al.*, 1982a). These results strongly suggest that the other component in the feedback loop is a rapidly turning over protein. Freeze-fracture studies where membrane proteins can be visualized showed that, while most organelle membrane proteins did not change over the circadian period, the protein complexes of the peripheral vesicle of *Gonyaulax* were both larger and more numerous during the night phase than during the day (Sweeney, 1976, 1981). This was true when cells were expressing their circadian rhythms in continuous light, perhaps merely another overt rhythm. However, the rapidly turning over membrane particles could just as well be transport proteins.

RNAS ARE RHYTHMIC

In *Gonyaulax,* the RNAs that can be extracted at different times of days are different, as shown by comparing their banding patterns on gels (Walz *et al.,* 1983). There is also evidence that the cells contain much more RNA just before midnight in LL, the concentration falling precipitiously just after midnight. Perhaps part of this RNA is translated into luciferase, which has recently been shown to be resynthesized in preparation for the night maximum in bioluminescence (Dunlap and Hastings, 1981; Johnson *et al.,* 1984).

THE CELL CYCLE IS NOT THE CLOCK

When eukaryotic cells divide mitotically, they traverse a definite cell division cycle. After a cell divides, it remains for various

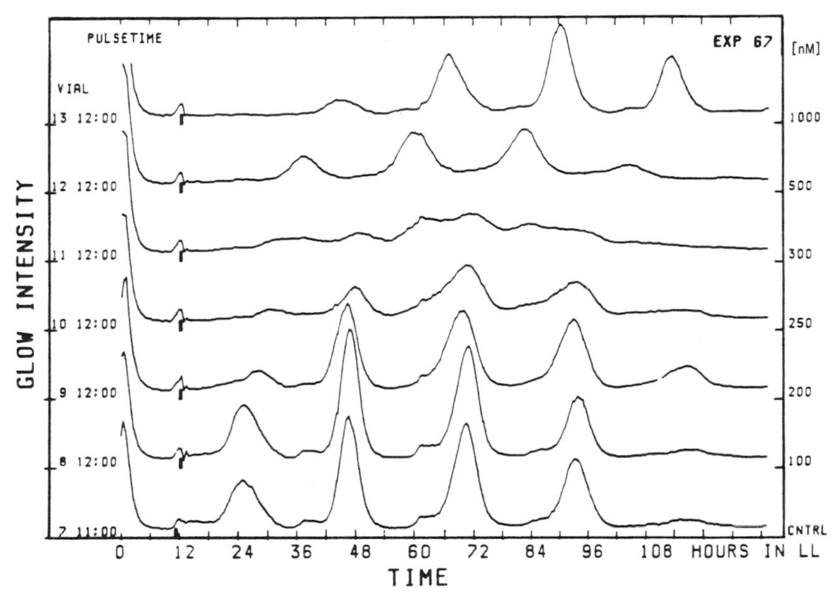

Fig. 4.4. The rhythm in the bioluminescence glow in *Gonyaulax polyedra* recorded with the "Taylortron." Cells in LL treated for 1 h at 12 h after transfer to LL with different concentrations of anisomycin as shown in nanomoles on the right ordinate. Note that the rhythm almost disappears when the cells are treated with 300 n*M* anisomycin at this time, but reappears when the anisomycin concentration is increased to 500 and 1000 n*M*, providing evidence for the presence of a singularity (Taylor, *et al.,* 1982a).

times in G_1 (gap 1) before beginning to synthesize more DNA for the new cell in the next stage, S for DNA synthesis. When synthesis of DNA is complete, the cell usually does not divide immediately, but passes through G_2 (gap 2) before initiating mitosis (M) and cytokinesis. Since as far as we know only eukaryotes show circadian rhythmicity, the question arose as to whether this eukaryotic cell division cycle might be the oscillator that times circadian rhythms. As we saw in Chapter 3, cell division in a number of dinoflagellates occurs at a characteristic time of day. The large dinoflagellate, *Pyrocystis fusiformis,* changes morphology five times during one cell cycle, which usually takes 4–5 days. By staining the nuclei with DAPI, a fluorescent stain specific for DNA, it was possible to discover which morphological stage corresponds to the S phase when the DNA doubles. Using these morphological stages as markers, we could show that cells of this species only progress from one part of the cell cycle to the next at defined circadian times. This led to the conclusion that the circadian clock controls the timing of the cell cycle rather than vice versa (Sweeney, 1982).

CONCLUSIONS ABOUT THE CLOCK MECHANISM FROM STUDIES WITH DINOFLAGELLATES

It is not possible at the present time to describe the circadian clock in dinoflagellates in any detail. It seems likely that there is only a single clock in a single cell. The clock is not the cell cycle. Ion transport across membranes and the synthesis of some protein or proteins seem to be important, perhaps intimate parts of the circadian clock.

A COMPARISON BETWEEN DINOFLAGELLATES AND OTHER RHYTHMIC ORGANISMS

As I mentioned at the beginning of this chapter, it is unwarranted at present to assume that the clocks controlling the timing of circadian rhythms are all alike. Are there strong similarities between the rhythmic behavior of different types of organisms or are there differences great enough to suggest an independent evolution for these phenomena? At the present time we do not know every detail of the mechanism of circadian timing in even a single organism. This question is in the minds of all who study the circadian rhythms. Is there evidence from other rhythmic organisms for the conclusions based on experiments with *Gonyaulax* and other dinoflagellates?

It is clear that the circadian rhythms in *Acetabularia* proceed normally in a single isolated cell, even in one without a nucleus (Sweeney and Haxo, 1961; Schweiger *et al.,* 1964; Mergenhagen and Schweiger, 1975). During vegetative growth, the nucleus of *Acetabularia* does not divide, thus, rhythmicity clearly does not depend on a cell cycle. A similar conclusion was reached by Edmunds and Adams (1981) from a somewhat more obscure line of reasoning.

ROLE OF IONS AND ION TRANSPORT IN TIME KEEPING

Although the circadian rhythm in the pulvinus of legumes depends on the coordinated changes in a number of cells, not just one, ion movement is certainly important in leaflet movements. Whether transport of ions is secondary, a "hand" of the clock, or an integral part of the clock is unclear. The ionophore, valinomycin, shifts the phase of the leaf movement rhythm in *Phaseolus* (Bünning and Moser, 1972). This experimental finding suggests that influx of the ion K^+, followed by water uptake, has a fundamental role in rhythmicity. An increase in the concentration of potassium can be detected in the extensor region during leaflet opening, both by scanning electron microscopy (SEM) with ion probe analysis on frozen sections of *Samanea* (Campbell and Garber, 1980; Satter and Galston, 1981) and by flame photometry in *Trifolium* (Scott and Gulline, 1975; Scott *et al.*, 1975). The latter authors also measured a rhythm in the rate of influx of $^{42}K^+$ in extensor (adaxial) cells of the clover pulvinus and a separate rhythm in membrane resting potential

(Scott *et al.*, 1977). Interestingly, these rhythms are not reset by valinomycin (Scott *et al.*, 1977). In fact, valinomycin is not universally active in rhythmic organisms.

In *Albizzia*, the rhythm is accompanied by changes in vacuolar structure, swollen cells having a single vacuole that becomes fragmented when the cell shrinks (Campbell and Garber, 1980). The volume of cytoplasm does not change during opening and closing of the leaflet. Scott *et al.* (1977) suggest that the long time needed for equilibration of ions throughout the tissue might account for the long circadian period. Thus, an organelle, the vacuole, is certainly responding to the clock but perhaps again only secondarily. As in *Gonyaulax*, the evidence points to ion transport in and out of organelles as a possible clock component.

The effects of exposure to other ions have been assayed in a variety of rhythmic organisms. Engelmann (1972) showed that the period of the rhythm of petal movement in *Kalanchoe* is increased in the presence of millimolar concentrations of lithium and that this is an effect directly on the light-sensitive oscillator (Engelmann *et al.*, 1974). The periods of some other rhythmic organisms, both plants and animals, are lengthened in the presence of lithium, including that of the growth rhythm of the diatom, *Skeletonema* (Östgaard *et al.*, 1982). Lithium has many diverse effects on cell physiology (Engelmann and Schrempf, 1980), so that conclusions from these experiments are uncertain. This article also contains an excellent summary of effects of ions and substances that could act on cellular ion concentration in circadian systems.

METABOLIC RHYTHMS AND TRANSPORT INTO AND OUT OF ORGANELLES

The rhythm in CO_2 fixation in *Bryophyllum* (see Chapter 3) can be almost completely explained as a feedback loop involving the activity of PEPcase and the transport of malate into and out of the vacuole (Fig. 4.5). It is known that the amount of this enzyme remains the same during the day and night. However,

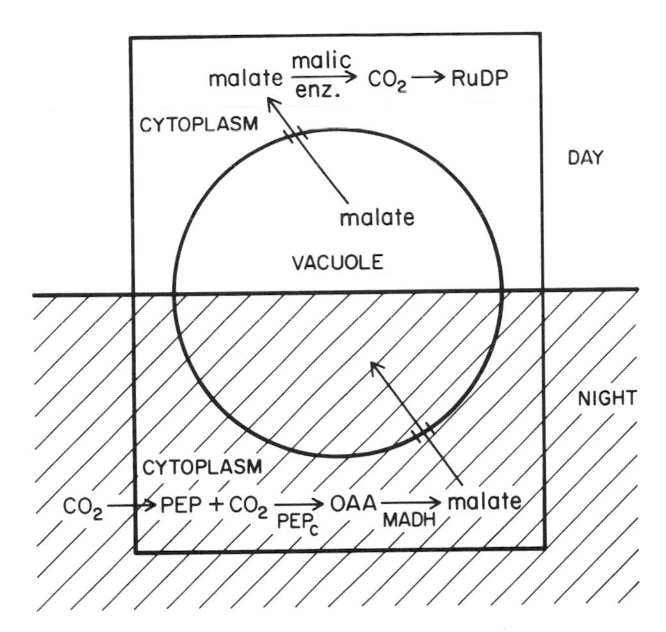

Fig. 4.5. Diagram of the biochemical feedback loop in *Bryophyllum* that may account for the circadian rhythm of CO_2 uptake.

the sensitivity to inhibition by the final product, malate, is 10 times less at night or during the night phase in DD as during the day. This change is caused by the phosphorylation of phosphoenolpyruvate carboxylase (PEPcase) at night (Nimmo *et al.*, 1984). However, what controls this timing of phosphorylation is not yet clear. During the night phase, the malate formed from oxaloacetate is accumulated in the vacuole, probably by a pump in the tonoplast. It is known that cycloheximide at low concentrations inhibits malate accumulation and the attendant acidification of the vacuole. The concentration of PEPcase is unaffected by this inhibitor of protein synthesis on 80S ribosomes (Bollig and Wilkins, 1979). These authors suggest that cycloheximide causes a change in the properties of the tonoplast. Perhaps the malate pump protein is not synthesized. During the day part of the circadian cycle, malate moves into the cytoplasm where it is broken down by malic enzyme. Why this takes place during the day and apparently not at night remains unexplained. It is clear however that a model very similar to that suggested for *Gonyaulax* could account for this rhythm, malate replacing potassium ions and the tonoplast being the membrane of importance.

Transport of an ion in and out of an organelle in *Euglena,* in this case the mitochondrion, has been suggested to be a feature of the oscillator in this unicell (Goto *et al.*, 1985). These workers proposed that NAD, NAD kinase, Ca^{2+}, and calmodulin form a biochemical feedback loop in *Euglena* and that calcium moves out of the mitochondria during part of this cycle (Fig. 4.6). In

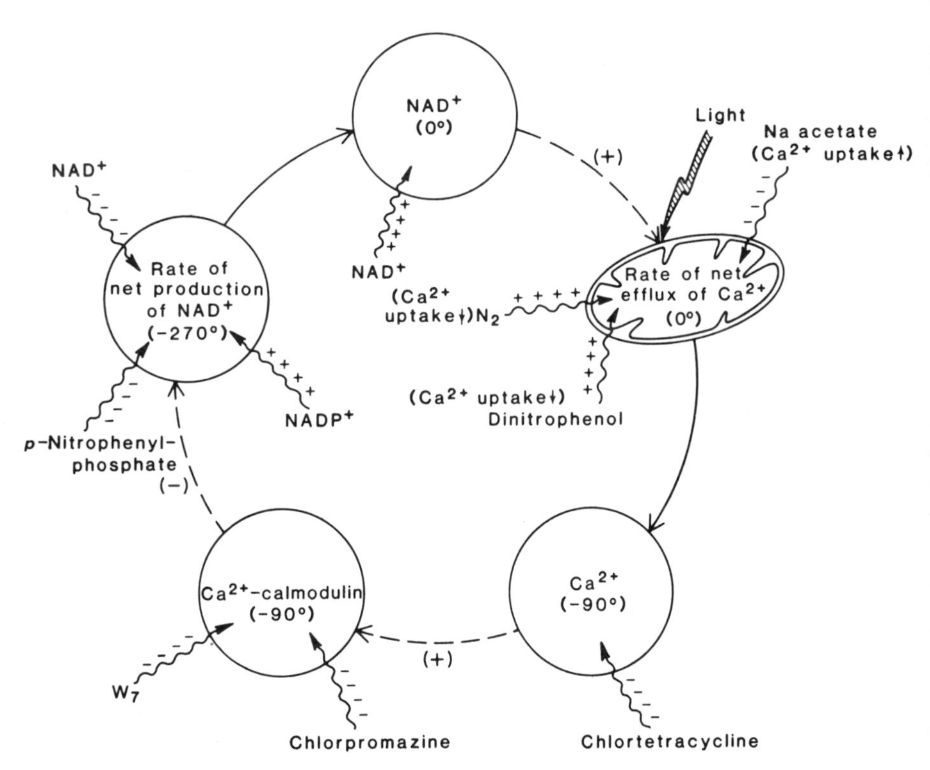

Fig. 4.6. Feedback loop proposed to explain the generation of circadian oscillations in *Euglena* and *Lemna* (Goto, *et al.,* 1985).

support of this hypothesis, rhythms in the activity of the enzymes involved have been demonstrated. Furthermore, both increasing and decreasing the amounts of substrates and calmodulin reset the rhythm in cell division and in opposite directions. A similar mechanism has been presented for the circadian rhythms in *Lemna gibba* (Goto, 1983).

THE ROLE OF PROTEIN SYNTHESIS IN THE CIRCADIAN OSCILLATOR

In *Acetabularia,* the rhythms that are known in photosynthesis—chloroplast shape and position and the amount of stored photosynthetic product—all originate in the chloroplast. The only possible exception is the rhythm in potential difference between the tip and the base of the stalk, the origin of which is obscure. Rhythms in this plant can be reset by cycloheximide, like the rhythm of bioluminescence in *Gonyaulax.* Furthermore, rifampicin, which inhibits translation on 70S ribosomes, has no effect on timing (Vanden Driessche *et al.,* 1970). Hence the protein important for circadian rhythmicity must be coded in the nucleus and translated in the cytoplasm rather than in the chloroplast. Schweiger and Schweiger (1977) have proposed a model for the circadian oscillator in *Acetabularia* that includes transport into the chloroplasts and the synthesis and degradation of a membrane protein that mediates this transport. Schweiger and his colleagues have detected several proteins that show circadian oscillations (Leong and Schweiger, 1979; Hartwig *et al.,* 1985). It will be most interesting to know what these proteins do in the cell. A model also hypothesizing protein synthesis and insertion into membranes under the influence of ion concentrations has been proposed by Burgoyne (1978).

Phase-shifting by substances that inhibit translation on 80S ribosomes in the cytoplasm of eukaryotes has been demonstrated in a number of organisms besides dinoflagellates and *Acetabularia,* among these *Bryophyllum* (Bollig and Wilkins, 1979), *Phaseolus* (Mayer and Knoll, 1981), and the sea hare,

Aplysia (Rothman and Strumwasser, 1976). The requirement for circadian protein synthesis is thus a common feature among rhythmic systems. We have no idea at present what proteins must be synthesized for proper clock function. There is one exception to this generalization: one of the *frq* mutants of *Neurospora* (see below), *frq 8*, is insensitive to cycloheximide (Feldman, 1982a). However, the protein important for clock function may be more stable than usual in this mutant.

MATHEMATICAL MODELS FOR CIRCADIAN RHYTHMS

Limit Cycles and the Singularity

While he was analyzing a large volume of data on the resetting behavior of *Drosophila pseudoobscura*, Winfree (1972) came upon a most interesting point until then unrecognized. He noticed that a light stimulus of a particular strength given at a unique time caused the fly population to lose rhythmicity in eclosion. A mathematician, Winfree interpreted this loss as evidence for a "singularity," a point in a limit cycle from which an oscillation can return to any position in the cycle. Such behavior in a population would result in asynchrony and thus no rhythm of the population would be measureable. The observation of a singularity thus was an argument for the participation of a limit cycle in circadian oscillations.

Singularities have indeed been observed in several plant rhythms as well as in *Drosophila*. *Kalanchoe* petal movement shows a singularity that can be detected in experiments on resetting the rhythm with light (Engelmann and Johnsson, 1978). Although *Gonyaulax* is too insensitive to phase-shifting by light to see the singularity in such experiments, it has been possible to detect the presence of a singularity in experiments where inhibitors of protein synthesis were used (Walz and Sweeney, 1979; Taylor *et al.*, 1982a). Rensing and Schill (1985) have been able to fit the phase responses of *Gonyaulax* to single pulses of anisomycin (Fig. 4.3) to a theoretical prediction from a limit cycle model. Certainly the existence of two types of PRCs, both with the same shape but one with a maximum near midnight, like PRCs to light pulses (Figs. 3.15 and 4.3), and the other with a maximum in the latter part of the day, like PRCs to dark pulses (Figs. 3.5 and 4.2), strongly suggest that there are two different components making up the clock oscillator as would be predicted by a limit cycle model. Thus, it is likely that a limit cycle may prove to be a part of the circadian clock.

THE CLOCK SHOP MODEL

The long period, about 24 h in circadian rhythms, has proven a stumbling block to devising a biochemical model for the oscillator, since metabolic reactions, protein synthesis, and even DNA transcription are all much too fast to generate such a long period. In consequence, it has been suggested that circadian oscillations are the result of beats between oscillations of much shorter periods (Pavlidis, 1971). Winfree (1975) has also considered that circadian time may be the result of averaging many independent circadian oscillators.

MUTANTS IN THE CIRCADIAN CLOCK: THE MOLD *NEUROSPORA CRASSA* AND THE FRUIT FLY *DROSOPHILA*

Mechanistic problems in biology have often been solved by a genetic approach using mutants. This strategy is now being applied to the problem of the nature of the circadian oscillator, including the powerful new tools of molecular biology. For these studies, plant systems have not yet proven tractable. Two organisms have been used so far for the molecular genetic approach, *Neurospora* and *Drosophila*. Both lend themselves to mutagenesis. *Neurospora* has many other advantages. This mold alternately grows vegetatively, then produces conidia which are dark in color. Hence, if spores are inoculated at one end of a trough containing solid medium, *Neurospora* writes its own record of

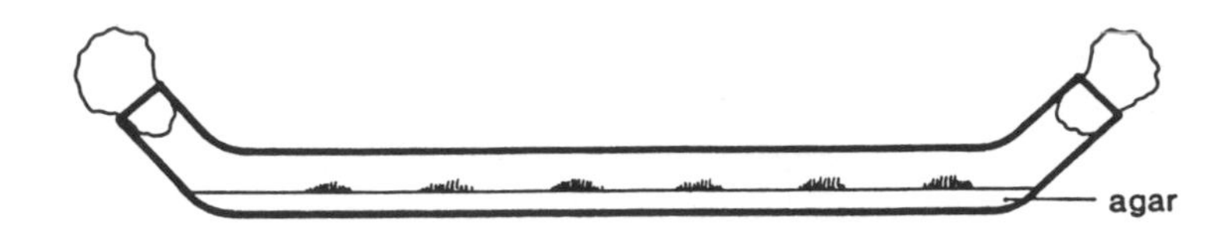

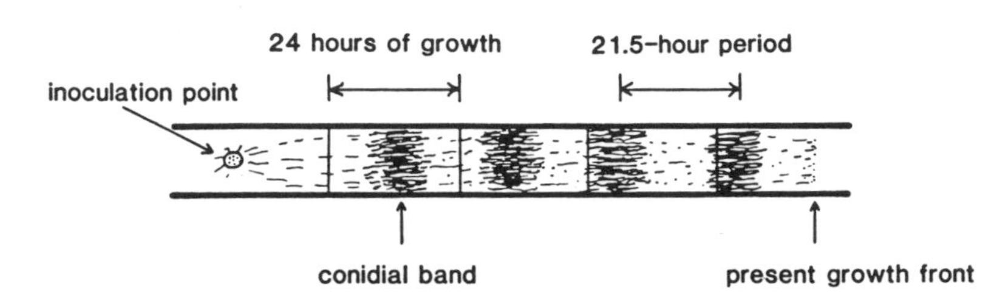

Fig. 4.7. Diagram of a culture of *Neurospora crassa*, band strain, growing in a race tube, seen from the side (above), and the top (below), showing the pattern of daily bands of conidia (Feldman, 1983).

rhythmicity as it grows and conidiates by turns (Fig. 4.7). Furthermore, much work has already been done on the genetics of this organism. A number of mutations that affect the rhythm have been isolated by Feldman and his collaborators, including eight at the same locus called *frq* or freak mutants (Feldman, 1982; 1983). Some of these have longer than normal periods while some have shorter periods. The techniques of molecular biology are being used to analyze how these mutations act and what their protein product is. This knowledge should shed much light on the nature of the circadian oscillator.

Rhythm mutants with periods longer or shorter than the wild type or without rhythmicity, the *per* mutants, have also been obtained in *Drosophila melanogaster*, a fruit fly which shows circadian rhythms in eclosion from the pupa and activity of the adult (Konopka and Benzer, 1971). The *per* locus has now been cloned and found to restore rhythmicity in arrhythmic mutants when the cloned DNA is injected into larvae (Bargiello *et al.*, 1984; Zehring *et al.*, 1984; Citri *et al.*, 1987). The DNA of the clone has been sequenced and it has an unusual number of repeats of the codons for threonine and glycine (Shin *et al.*, 1985). The sequence of the open reading frame in the cloned DNA resembles that of genes coding for peptidoglycans (Jackson *et al.*, 1986). When the nature of the gene product and its physiological function is completely known, an exciting insight into the circadian clock is sure to result. Other DNAs have been probed with the plasmid containing the cloned *per* locus and there is evidence for homology with mouse, cat, and human DNA but not with yeast (Shin *et al.*, 1985), an organism in which no circadian rhythms are currently known.

GENERAL CONCLUSIONS

In general, then, studies of the mechanism for circadian rhythmicity using different organisms, as different as algae and mammals, have not made it necessary to postulate distinctive mechanisms arising by independent evolution. Perhaps we simply do not yet know enough to make this distinction. The work on genes that affect timing, particularly the cloning of these genes and the identification of the gene products and their functions, offer hope of more understanding soon. It seems that some vital piece of the puzzle is still missing. Perhaps the most likely place to find this piece is among the mechanisms for biological control, at the level of DNA, membrane signal transduction, or enzyme activity. Research in all these areas is now very active and each new bit of information will be assayed for a role in the biological clock, as has been done in the past. We desperately need a more reliable way to distinguish experimentally between the oscillations that are part of the clock and those that are "hands," merely responding to the clock output. Although we have made progress, I think that much more research will be necessary before we understand the mechanism that generates these fascinating circadian rhythms.

REFERENCES

Adamich, M., Laris, P. C., and Sweeney, B. M. (1976). *In vivo* evidence for a circadian rhythm in membranes of *Gonyaulax. Nature (London)* **261**, 583–585.

Bargiello, T., Jackson, R., and Young, M. (1984). Restoration of circadian behavioral rhythms by gene transfer in *Drosophila. Nature (London)* **312**, 752–754.

Bollig, I. C., and Wilkins, M. B. (1979). Inhibition of the circadian rhythm of CO^2 metabolism in *Bryophyllum* leaves by cycloheximide and dinitrophenol. *Planta* **145**, 105–112.

Brown, F. A., Jr., Hastings, J. W., and Palmer, J. D. (1970). "The Biological Clock, Two Views." Academic Press, New York.

Bünning, E., and Moser, I. (1972). Influence of valinomycin on circadian leaf movements of *Phaseolus. Proc. Natl. Acad. Sci. U. S. A.* **69**, 2732–2733.

Burgoyne, R. D. (1978). A model for the molecular basis of circadian rhythms involving monovalent ion-mediated translational control. *FEBS Lett.* **94**, 17–19.

Campbell, N. A., and Garber, R. C. (1980). Vacuolar reorganization in the motor cells of *Albizzia* during leaf movement. *Planta* **148**, 251–255.

Chay, T. R. (1981). A model for biological oscillations. *Proc. Natl. Acad. Sci. U. S. A.* **78**, 2204–2207.

Citri, Y., Colot, H. V., Jacquier, A. G., Yu, Q., Hall, J. C., Baltimore, D., and Rosbash, M. (1987). A family of unusually spliced biologically active transcripts encoded by a *Drosophila* clock gene. *Nature (London)* **326**, 42–47.

Dunlap, J. C., and Hastings, J. W. (1981). The biological clock in *Gonyaulax* controls luciferase activity by regulating turnover. *J. Biol. Chem.* **141**, 1269–1270.

Dunlap, J. C., Taylor, W., and Hastings, J. W. (1980). The effects of protein synthesis inhibitors on the *Gonyaulax* clock. I. Phase-shifting effects of cycloheximide. *J. Comp. Physiol.* **138**, 1–8.

Edmunds, L. N., Jr., and Adams, K. J. (1981). Clocked cell cycle clocks. *Science* **211**, 1002–1013.

Engelmann, W. (1972). Lithium slows down the *Kalanchoe* clock. *Z. Naturforsch., B: Anorg. Chem., Org. Chem., Biochem., Biophys., Biol.* **27B**, 477.

Engelmann, W., and Johnsson, A. (1978). Attenuation of the petal movement rhythm in *Kalanchoe* with light pulses. *Physiol. Plant.* **43**, 68–76.

Engelmann, W., and Schrempf, M. (1980). Membrane models for circadian rhythms. *Photochem. Photobiol. Rev.* **5**, 49–86.

Engelmann, W., Maurer, A., Muhlbach, M., and Johnsson, A. (1974). Action of lithium ions and heavy water in slowing circadian rhythms of petal movement in *Kalanchoe. J. Interdiscip. Cycle Res.* **5**, 199–205.

Feldman, J. F. (1982). Genetic approaches to circadian clocks. *Annu. Rev. Plant Physiol.* **33**, 583–608.

Feldman, J. F. (1983). Genetics of circadian clocks. *Bioscience* **33**, 426–431.

Goto, K. (1983). Causal relationships among metabolic circadian rhythms in *Lemna. Z. Naturforsch., C: Biosci.* **39C**, 73–84.

Goto, K., Laval-Martin, D. L., and Edmunds, L. N., Jr. (1985). Biochemical modeling of an autonomously oscillatory circadian clock in *Euglena. Science* **228**, 1284–1288.

Hamner, K. C., Finn, J. C., Sirohi, G. S., Hoshizake, T., and Carpenter, B. H., (1962). The biological clock at the South Pole. *Nature (London)* **195**, 476–480.

Hartwig, R., Schweiger, M., Schweiger, R., and Schweiger, H. G. (1985). Identification of a high molecular weight polypeptide that may be part of the circadian clockwork in *Acetabularia. Proc. Natl. Acad. Sci. U. S. A.* **82**, 6899–6902.

Hastings, J. W. (1960). Biochemical aspects of rhythms: Phase shifting by chemicals. *Cold Spring Harbor Symp. Quant. Biol.* **25**, 131–143.

Hastings, J. W., and Schweiger, H.-G., eds. (1976). "The Molecular Basis of Circadian Rhythms," *Life Sci. Res. Rep.* No. 1. Dahlem Konferenzen, Berlin.

Jackson, F. R., Bargiello, T. A., Yun, S.-H., and Young, M. W. (1986). Product of the *per* locus of *Drosophila* shares homology with proteoglycans. *Nature (London)* **320**, 185–188.

Johnson, C. H., Roeber, J. F., and Hastings, J. W. (1984). Circadian changes in enzyme concentration account for rhythm of enzyme activity in *Gonyaulax. Science* **223**, 1428–1430.

Karakashian, M. W., and Hastings, J. W. (1962). The inhibition of a biological clock by actinomycin D. *Proc. Natl. Acad. Sci. U. S. A.* **48**, 2130–2137.

Karakashian, M. W., and Hastings, J. W. (1963). The effects of inhibitors of macromolecular biosynthesis upon the persistent rhythm of luminescence in *Gonyaulax. J. Gen. Physiol.* **47**, 1–12.

Karakashian M. W., and Schweiger, H. G. (1976). Evidence for a cycloheximide-sensitive component in the biological clock of *Acetabularia. Exp. Cell Res.* **98**, 303–312.

Kiessig, R. S., Herz, J. M., and Sweeney, B. M. (1979). Shifting the phase of the circadian rhythm in bioluminescence in *Gonyaulax* with vanillic acid. *Plant Physiol.* **63**, 324–327.

Konopka, R. J., and Benzer, S. (1971). Clock mutants of *Drosophila melanogaster. Proc. Natl. Acad. Sci. U. S. A.* **68**, 2112–2116.

Krasnow, R., Dunlap, J., Taylor, W., and Hastings, J. W. (1981). Measurements of *Gonyaulax* bioluminescence, including that of single cells. *In* "Bioluminescence, Current Prospectives" (K. H. Nealson, ed.), pp. 52–63. Burgess, Minneapolis, Minnesota.

Leong, T.-Y., and Schweiger, H.-G. (1979). The role of chloroplast-membrane-protein synthesis in the circadian clock. *Eur. J. Biochem.* **98**, 187–194.

McMurry, L., and Hastings, J. W. (1972). No desynchronization among four circadian rhythms in the unicellular alga, *Gonyaulax polyedra. Science* **175**, 1137–1139.

Mayer, W. E., and Knoll, U. (1981). Temperature compensation of cycloheximide-sensitive phases of the circadian clock in the *Phaseolus* pulvinus. *Z. Pflanzenphysiol.* **103**, 413–425.

Mergenhagen, D., and Schweiger, H.-G. (1975). The effect of different inhibitors of transcription and translation on the expression and control of circadian rhythm in individual cells of *Acetabularia. Exp. Cell Res.* **94**, 321–326.

Nimmo, G. A., Nimmo, H. G., Fewson, C. A., and Wilkins, M. B. (1984). Diurnal changes in the properties of phosphoenolpyruvate carboxylase in *Bryophyllum* leaves: A possible covalent modification. *FEBS Lett.* **178**, 199–203.

Njus, D., Sulzman, F. M., and Hastings, J. W. (1974). Membrane model for a circadian clock. *Nature (London)* **248**, 116–120.

Ostgaard, K., Jensen, A., and Johnsson, A. (1982). Lithium ions lengthen the circadian period of growing cultures of the diatom *Skeletonema costatum. Physiol. Plant.* **55**, 285–288.

Pavlidis, T. (1971). Populations of biochemical oscillators as circadian clocks. *J. Theor. Biol.* **33**, 319–338.

Rensing, L., and Schill, W. (1985). Perturbation by single and double pulses as analytical tool for analysing oscillatory mechanisms. *In* "Tempoaral Order" (L. Rensing and N. Jaeger, eds.), pp. 226–231. Springer-Verlag, Berlin and New York.

Rothman, B. S., and Strumwasser, F. (1976). Phase shifting the circadian rhythm of neuronal activity in the isolated *Aplysia* eye with puromycin and cycloheximide: Electrophysiological and biochemical studies. *J. Gen. Physiol.* **68**, 359–384.

Satter, R. L., and Galston, A. W. (1981). Mechanisms of control of leaf movements. *Annu. Rev. Plant Physiol.* **32**, 83–110.

Schweiger, H.-G., and Schweiger, M. (1977). Circadian rhythms in unicellular organisms: an endeavor to explain the molecular mechanism. *Int. Rev. Cytol.* **51**, 315–342.

Schweiger, E., Wallraff, H. G., and Schweiger, H. G. (1964). Uber tagesperiodisches Schwankungen der Sauerstoffbilanz kernhaltiger und kernloser *Acetabularia mediterranea. Z. Naturforsch.* **19B**, 499–505.

Scott, B. I. H., and Gulline, H. (1975). Membrane changes in a circadian system. *Nature (London)* **254**, 69–70.

Scott, B. I. H., Gulline, H., and Robinson, G. R. (1975). Circadian electrical and ionic changes at the membranes of pulvinal cells in *Trifolium repens. Plant Physiol.* **56**, 76.

Scott, B. I. H., Gulline, H. F., and Robinson, G. R. (1977). Circadian electrochemical changes in the pulvinus of *Trifolium repens* L. *Aust. J. Plant Physiol.* **4**, 193–206.

Shin, H.-S., Bargiello, T. A., Clark, B. T., Jackson, F. R., and Young, M. W. (1985). An unusual coding sequence from a *Drosophila* clock gene is conserved in vertebrates. *Nature (London)* **317**, 445–448.

Sulzman, F. M., Ellman, D., Fuller, C. A., Moore-Ede, M. C., and Wassmer, G. (1984). *Neurospora* circadian rhythms in space: A reexamination of the endogenous-exogenous question. *Science* **225**, 232–234.

Sweeney, B. M. (1960). The photosynthetic rhythm in a single cell of *Gonyaulax polyedra. Cold Spring Harbor Symp. Quant. Biol.* **25**, 145–148.

Sweeney, B. M. (1963). Resetting the biological clock in *Gonyaulax* with ultraviolet light. *Plant Physiol.* **38**, 704–708.

Sweeney, B. M. (1974a). The potassium content *Gonyaulax polyedra* and phase changes in the circadian rhythm of stimulated bioluminescence by short exposures to ethanol and valinomycin. *Plant Physiol.* **53**, 337–342.

Sweeney, B. M. (1974b). A physiological model for circadian rhythms derived from the *Acetabularia* rhythm paradoxes. *Int. J. Chronobiol.* **2**, 25–33.

Sweeney, B. M. (1976a). Freeze-fracture studies of *Gonyaulax polyedra.* I. Membranes associated with the theca and circadian changes in the particles of one membrane face. *J. Cell Biol.* **68**, 451–461.

Sweeney, B. M. (1976b). Evidence that membranes are components of circadian oscillators. *In* "The Molecular Basis of Circadian Rhythms" (J. W. Hastings and H.-G. Schweiger, eds.), pp. 267–281. Dahlem Konferenzen, Berlin.

Sweeney, B. M. (1981). Freeze-fractured chloroplast membranes of *Gonyaulax polyedra* (Pyrrophyta). *J. Phycol.* **17**, 95–101.

Sweeney, B. M. (1982). Interaction of the circadian cycle with the cell cycle in *Pyrocystis fusiformis. Plant Physiol.* **70**, 272–276.

Sweeney, B. M., and Haxo, F. T. (1961). Persistence of a photosynthetic rhythm in enucleated *Acetabularia. Science* **134**, 1361–1363.

Sweeney, B. M., and Herz, J. M. (1977). Evidence that membranes play an important role in circadian rhythms. *Proc. Int. Conf. Int. Soc. Chronobiol., 12th,* pp. 751–761.

Taylor, W., and Hastings, J. W. (1979). Aldehydes phase shift the *Gonyaulax* clock. *J. Comp. Physiol. B.* **130**, 359–362.

Taylor, W., Gooch, D. van, and Hastings, J. W. (1979). Period-shortening and phase-shifting effects of ethanol on the *Gonyaulax* glow rhythm. *J. Comp. Physiol. B.* **130**, 355–358.

Taylor, W., Krasnow, R., Dunlap, J. C., Broda, H., and Hastings, J. W. (1982a). Critical pulses of anisomycin drive the circadian oscillator in *Gonyaulax* toward its singularity. *J. Comp. Physiol.* **148**, 11–25.

Taylor, W., Krasnow, R., Dunlap, J. C., and Hastings, J. W. (1982b). Inhibi-

Rhythms That Match Environmental Periodicities: Tidal, Semilunar, and Lunar Cycles

Along the edges of the ocean, especially where there is a large tidal difference or a very gradual slope, the alternation of high and low tide is perhaps the most spectacular feature of the environment. Organisms that live in the intertidal zone are alternately exposed and under water. The attendant changes in temperature and light intensity may be large. It is not surprising then that many of these organisms show tidal rhythms.

TIDAL RHYTHMS

The tidal cycle is the result of the interaction of the gravity of the sun and the moon on the earth. While both water and dry land are affected, the tides in large bodies of water are much the larger. The fundamental period of the tidal cycle is 24.8 h, and is often subdivided into a major and minor tide, 12.4 h apart. The period of a full tidal cycle is thus only slightly different from a circadian cycle. However, there is an important difference between a true tidal and a circadian rhythm. Because the tidal cycle is slightly longer than a day, high tide and low tide occur later each day by almost 1 h. This means that to be effective, a tidal rhythm *cannot* be reset by light. A true tidal rhythm therefore cannot be simply a circadian rhythm with a long period.

The clearest and best known tidal rhythms are those found in invertebrates that live on or close to the shore. The flatworm *Convoluta roscoffensis*, which contains symbiotic green algae, emerges from the sand only at low tide. Along the northwestern coast of France, the beaches can become green as the water recedes because of the large numbers of *Convoluta* (Bohn, 1903). This worm with its symbionts can now be cultured in the laboratory (Provasoli *et al.*, 1968).

Another invertebrate, the amphipod *Synchelidium*, leaves the sand only when the water covers its beach habitat. Enright (1963) has shown that, when it is brought into the laboratory and kept in darkness in vessels containing seawater and sand, *Synchelidium* continues its tidal cycle of swimming and resting. In the laboratory, it mimics quite accurately not only the tidal cycle but even the details of the tidal fluctuations on its native beach. Thus, the tidal behavior of *Synchelidium* constitutes a true tidal rhythm, and does not depend on the animal's being directly in contact with the tide.

Another interesting example of a tidal pattern was found by Rao (1954) in mussels. These molluscs pump more water through their gills at times when the tide is high than during low tide. When the amount of seawater passing through the animal is measured in the laboratory in the absence of any tidal differences, the pumping rate is found to vary, rising and falling in almost perfect synchrony with the water level at the site from which the mussels were collected, even mirroring the variations in the magnitude of the tidal ebb and flow quite accurately, as does *Synchelidium*. Furthermore, no temperature dependence in pumping rate between 9 °C and 20 °C could be found and the tidal rhythm persisted in both constant light and constant darkness. The same rhythm was found in mussels collected from a depth of 30 m and those from 1 m. Animals growing on floats, and hence not subject to any differences in water level, still followed the same tidal rhythm as animals growing on nearby pier pilings. Such observations implied that the rhythm was not under the direct control of the tidal water level but constituted an endogenous cycle synchronized to the tide in some other way. That these cycles were indeed endogenous was confirmed by the observation that mussels collected on the east coast at Woods Hole, Massachusetts, and flown to California showed a tidal pumping rhythm in the laboratory that continued in time with the tide in their east coast home, out of synchrony with the west coast tidal changes. When these mussels were exposed to the tide by hanging in baskets from the pilings of a pier in Corona del Mar, California, however, their tidal pumping rhythm was quickly reset to match their west coast environment. Therefore there must be some entraining agent associated with the tides. Both *Mytilus* and *Synchelidium* ape the local tidal changes so closely that the tide itself rather than the moon would seem to be the determining factor.

For many years it has been noted that the mud of tidal flats changes color shortly after it is exposed as the tide falls. The color change is brought about by the accumulation at the surface of very large numbers of motile pigmented algae. Fauvel and Bohn (1907) were perhaps the first to investigate whether or not such an alga would continue to appear on the surface of mud samples brought to the laboratory and hence no longer under the direct influence of the tide. They found that the diatom *Pleurosigma* did alternately accumulate and disappear in the mud samples for some time, and seemed to appear later each day, as did the tide.

Another example of this phenomenon was studied by Bracher in 1937. The mouth of the River Avon in Bristol, England, is an estuary with very large tidal differences in water level between high and low tide. The mud flats that are exposed at low tide become green in spots due to the accumulation of *Euglena obtusa* (called by Bracher *E. limosa*). She reported that when she brought the organisms to the laboratory, they continued to move up and down with a tidal period. However, Palmer and Round (1965) have reinvestigated the movements of this *Euglena* from the River Avon. They confirmed the field observations of Bracher, using small pieces of lens tissue laid on the surface of the mud to trap the organisms there for counting. They also made miniature mud cores, which they froze and sectioned while still frozen to determine where the organisms were found at different times in the tidal cycle. *Euglena* cells did accumulate at the surface during low tides which occurred during daylight. Before the tide came in again, they moved down into the mud (Fig. 5.1, upper curve). None were found at the surface when low tide occurred at night. This would not be surprising if motion upward always consisted of swimming toward the light. Laboratory experiments revealed an unexpected fact: as soon as this *Euglena* was brought into the laboratory and placed in LL at 980 lux, the

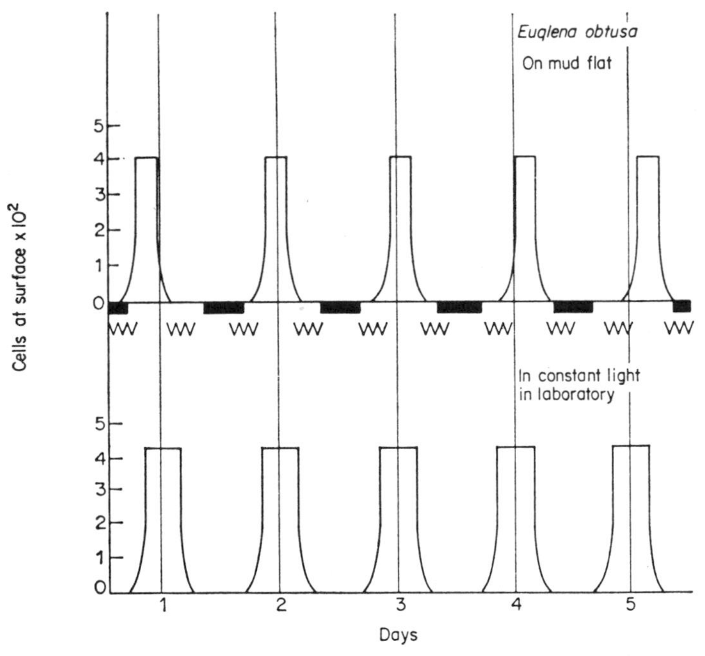

Fig. 5.1. The vertical migration rhythm in *Euglena obtusa* on a mud flat (top curve), and in constant light in the laboratory (bottom curve). The dark bars on the upper abscissa represent night, while the wavy lines show when the tide covered the mud flat (Palmer and Round, 1965).

cells displayed a typical *circadian* rhythm (Fig. 5.1, lower curve). In constant light, the period was approximately the same between 5 and 18.5 °C, the Q_{10} being about 0.9. The phase of this circadian rhythm was determined by the last dark period in nature, whether it was a night or a dark period resulting from the covering of the mud flat by murky water at high tide. Thus, the rhythm was actually circadian, entrained by the tidally caused light–dark cycle, rather than truly tidal. Round, Palmer, and Happey also investigated a number of diatoms from the River Avon mud flats and these too were shown to possess circadian, rather than tidal endogenous, rhythms (Round and Happey, 1965; Round and Palmer, 1966). The question remained whether possibly all apparently tidal cycles in plants were merely manifestations of circadian rhythms reset by light intensity differences associated with the incoming tide.

Surirella gemma is one of the mudflat diatoms that in nature collects at the surface at low tide (Hopkins, 1966a,b). It too reverts to a circadian behavior in the laboratory in a light–dark cycle and the rhythmicity persists with a circadian period in LL. In this species the resetting stimulus is the wetting of the cells as the tide comes in.

Fauré-Fremiet (1950) made observations of a Chrysomonad, *Chromulina*, which, like the diatoms and *Euglena*, moves up at low tide and colors the mud flats of its habitat. He found that in this organism the direction of phototaxis clearly reverses, being positive at low tide and negative at high tide. The presence of mud was not necessary for the observation of reversal in phototaxis, which continued to occur in the laboratory for about 6–7 days after collection, more or less synchronized with the tide. The changes in the direction of phototaxis gradually became less marked until cells were always positively phototactic. It is possible that rhythms in phototaxis also underlie the tidal rhythms in diatoms such as *Surirella*.

Fauré-Fremiet (1951) later visited Woods Hole and while there he noted that brown spots appeared at low tide on the mud flats of Barnstable Harbor nearby. The diatom *Hantzschia virgata* proved to be responsible. This diatom can glide half its length in a second. Cells examined in a drop of water were positively phototactic at low tide. At high tide they moved away from the light and seemed to exude a mucilaginous material, which caused them to stick to each other and to sand grains. The phototactic rhythm disappeared after a few days in the laboratory.

After finding that the "tidal" rhythm in *Euglena obtusa* and the diatoms of the River Avon was in reality a typical circadian rhythm reset by the murky water at high tide, Palmer and Round were interested to see whether this was also the case with *Hantzschia*, which grew in clearer waters. Consequently they reinvestigated the behavior of *Hantzschia* collected from the same Barnstable Harbor where Fauré-Fremiet had found them and brought them into the laboratory at Woods Hole. Cells at the mud surface were counted as before by the lens tissue technique. Unlike the diatoms previously studied, *Hantzschia* still showed a clearly tidal rhythm in the laboratory, migrating up at intervals of 24.8 h in continuous light (Fig. 5.2) and even in a 12:12 light–dark cycle. Circadian rhythms never show periods other than 24 h when entrained by a 24-h light–dark cycle. Another striking observation pointed to the presence in *Hantzschia* of a rhythm other than a modified circadian one. In LD, no migration took place during the dark period, and when cells were transferred to LL, still no migration occurred in what corresponded to the night phase of a circadian rhythm. The maximum accumulation of cells at the mud surface was later each day in accordance with a tidal cycle until the night phase was reached. In the next cycle no migration was seen at all (Fig. 5.2), but on the following day a peak appeared in the morning. The interpretation offered by Palmer (1976) for this observation was that the true period of the tidal rhythm was 12.4 h but the tidal maximum could not be expressed when it coincided with the night phase of the circadian rhythm. Such a situation would result from the interaction of a circadian rhythm in the ability to move and a tidal rhythm in the direction of movement. Little is known of these underlying rhythms in *Hantzschia* since all efforts to culture this diatom have failed. Further work on the subject of tidal rhythms in diatoms and euglenids might provide very interesting results.

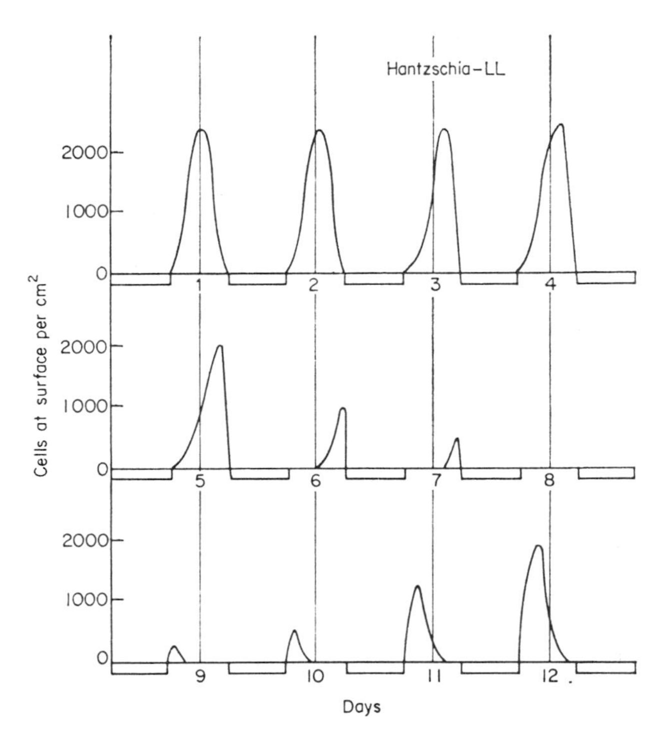

Fig. 5.2. The vertical-migration rhythm in the diatom *Hantzschia* during 12 days in the laboratory under constant illumination. The bars under the abscissa represent the night phase of a postulated circadian rhythm (Palmer and Round, 1967).

A tidal component has been postulated for the photosynthesis of the macroalga *Dictyota dichotoma* (Vielhaben, 1963). Oxygen evolution from the thallus of this brown alga varies during the course of a day, but the time of the apparent maximum changes, being later on each successive day, in accord with a 24.6-h period. The variation appears to be caused by interaction between a tidal and a diurnal rhythm in which the tidal effect is inhibitory, taking a progressively later bite out of the diurnal maximum each day, as if a shadow of the moon were eclipsing the daily photosynthesis. The importance of this tidal rhythm will be discussed below as part of the interpretation offered by

Müller (1962) and Vielhaben (1963) for the semilunar rhythm in reproduction that they demonstrated in *Dictyota.*

True tidal rhythms, like circadian rhythms, require some way to be synchronized with the environment. It is not clear what the cues are. That some cue does exist is evident, because mussels transported from the east coast to the west coast of North America and attached to a pier piling soon become resynchronized to the new local tides. Mussels suspended from a floating buoy, however, did not change the timing of their tidal rhythms, an observation which suggests that wave action may be the cue. Water turbulence also resets the tidal rhythm in the amphipod *Synchelidium* and the isopod *Excirolana chiltoni.* The phase-response curve (PRC) for 2-h exposures to simulated waves in *Excirolana* turned out to show two peaks in 25 h, quite unlike any PRCs for circadian rhythms (Enright, 1976). Perhaps, however, there are a number of signals such as changes in pressure, water motion, or drying and wetting that vary with the tides and serve to synchronize tidal rhythms with the environment.

SEMILUNAR AND LUNAR RHYTHMS

The tides are caused by the effects of the gravity of the sun and moon on the earth. Still another environmental periodicity results from the relative motion of earth, moon, and sun. A new moon, then a full moon, appear once in each lunar month of 29.5 days. Twice each lunar month, or every 14.8 days, the tidal changes in water level are more extreme than usual. This occurs at new and full moon when the gravity of the sun and the moon are pulling along the same line, and create a higher tide, called the "spring tide." When the moon is in its first or third quarter and the sun and moon pull at right angles to each other, the tidal range is least. Such tides are known as "neap tides." Are there any periodicities in plants or animals which match these lunar and semilunar cycles?

Very little is known concerning even the existence of rhythmicities with periods that match either the lunar month or the semilunar spring and neap tides. There are, however, among marine animals a few examples that make up for their rarity by their spectacular nature. Perhaps the most famous example concerns the behavior of the grunion, a small fish that inhabits the sea off California. This fish spawns on the nights just after the highest spring tide twice each lunar month in the spring with quite remarkable predictibility (Walker, 1952): the expected night and time are printed in the local tide tables.

During spawning, this lithe little fish actually leaves the water and wiggles up on the beach to lay eggs at the level reached only by the highest wave or to fertilize the eggs after they are laid. The fish swim up in one wave, lay or fertilize the eggs, and return to the sea on the following wave. The embryos develop for 15 days in the sand above the reach of the waves. They do not leave the egg until wet by the high water of the next spring tide 2 weeks later. This behavior and its timing cannot so far be studied under controlled conditions, consequently it is not known whether an endogenous cycle cues the grunion when it is time to spawn.

Another remarkable semilunar periodicity can be observed in the Palolo worm, *Eunice*, a polychaete that is found off Bermuda, in the East Indies and Pacific islands (Hauenschild *et al.*, 1968). This worm mates only just after sunset at the dark of the moon. The female comes to the surface, swims in a circle, and emits a brilliantly bioluminescent secretion containing the eggs. The male is attracted and fertilizes the eggs while also emitting light. In the Indonesian species whose mating display I have had the good fortune to see, only the rear part of the worm containing the reproductive organs comes to the sea surface to mate. It even has a false head. The rest of the worm stays safely on the bottom of the sea. Another marine polychaete *Platynereis dumerilii* also regulates its reproductive cycles to match the phase of the moon. This organism has been studied under laboratory conditions by Hauenschild (1960). Metamorphosis into the sexually capable form is followed by swarming of the reproductive parts, which separate from the rest of the animal as in the Palolo worm. Swarming occurs at the dark of the moon. Under a 12:12 LD however, swarming is random. It can be synchronized by 6 days of constant light every 30 days, the time of swarming being determined by the end of the light period. A very weak light at night is sufficient to cause lunar periodicity in swarming of *Playnereis.* After one exposure to light at night in the laboratory, semilunar spawning can continue for a number of cycles if the experiment is started when the animals are young. Thus, this worm appears to be capable of an endogenous lunar rhythm.

The marine midge *Clunio marinus* has an interesting semilunar cycle, the insect eclosing only at the very low tides of the spring series and only during the afternoon (Neumann, 1976). In nature and in the laboratory, this behavior is controlled by real or artificial moonlight at night and by turbulence of real or simulated low tide. After being entrained by exposure to weak light (0.4 lux) during four successive nights, the semilunar rhythm of eclosion was observed for at least two cycles, and hence appeared to be endogenously controlled.

These examples of semilunar rhythms in animals allow them to take advantage of the extremely high or low tides that accompany the spring tide series. Are there comparable adaptations in plants of the intertidal zone? Of course there are. A number of attached algae, including the green algae *Derbesia* (Ziegler-Page and Kingsbury, 1968), *Ulva* (Smith, 1947), *Enteromorpha* (Christie and Evans, 1962), and the brown alga *Dictyota* (Hoyt, 1927; Bünning and Müller, 1961), discharge gametes only at the very low tides so that the reproductive cells are less diluted when they are discharged into the water. Two of these species have been studied under controlled conditions in the laboratory. Müller (1962) was able to grow *D. dichotoma* in culture and obtain fruiting, which he measured by counting the number of eggs released. When pieces of the thallus were cultivated under natural light, gametes were produced in bursts that occurred every 14–15 days. However, when the plants were moved to an artificial light–dark cycle (LD 14:10), few eggs were released and no synchrony could be seen. An explanation of these apparently contradictory findings was suggested by the experiments of Hauenschild: perhaps as in *Platynereis*, light at night, such as moonlight under natural conditions, may be necessary to synchronize the production of gametes. Müller tested this hypothesis by leaving the lights on in the laboratory during one night, thus exposing his *Dictyota* to a simulated moon. Ten days later a burst of egg production was indeed observed. A cycle in reproduction continued to be expressed in which a maximum occurred every 16 days thereafter for at least five semilunar cycles (Fig. 5.3). At night an intensity of light of 3 lux was enough to produce this effect.

To account for the semilunar period of the observed rhythm, Müller (1962) suggested that both a tidal and a diurnal rhythm were present in *Dictyota* and that these two periodicities interacted. When a period of 24 h and a period of 12.4 h are simultaneously affecting the same physiological process, the maxima of these two periodicities will coincide at intervals of about 15 days.

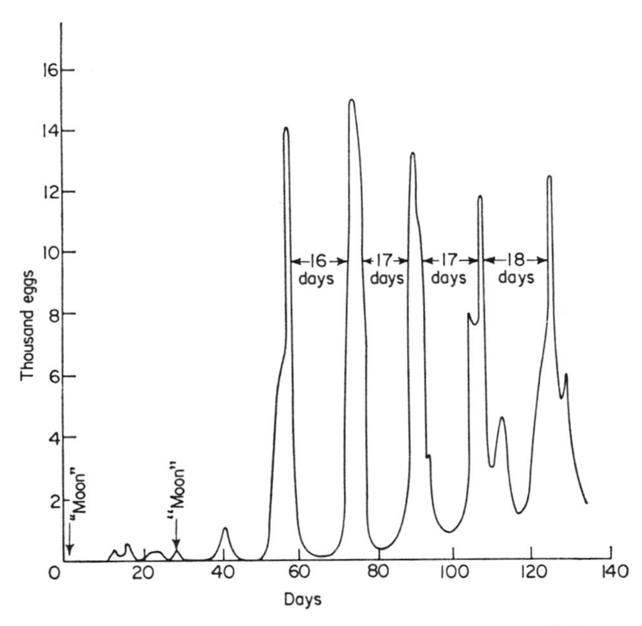

Fig. 5.3. Egg production by the brown alga *Dictyota dichotoma* in the laboratory growing with a 14:10 LD cycle, 20°C. The dark period was replaced by light on day 1 and day 28, as shown by the arrows on the abscissa, to give the plants artificial 'moonlight' (Müller, 1962).

If the resulting reinforcement or "beat" could conceivably cue a response, then this response will show a period of about 15 days, a semilunar periodicity. This hypothesis can be tested by changing the frequency of one of these rhythms and consequently changing the timing of the beat. The way in which tidal rhythms can be reset in plants is unknown, but circadian rhythms can easily be entrained to a different period by changing the LD cycle of the environment. Müller began a series of experiments in which *Dictyota* was maintained on a 23.5-h light–dark regime, a schedule that theoretically should result in a beat every 12 days instead of every 15. He did indeed find a shortening of the next interval between one bout of egg production and the next, but only to 13–14 days. His experiments were extended by Vielhaben (1963), who examined fruiting of *Dictyota* on 23- and 24.5-h LD cycles (Fig. 5.4). As before, shortening the day shortened the

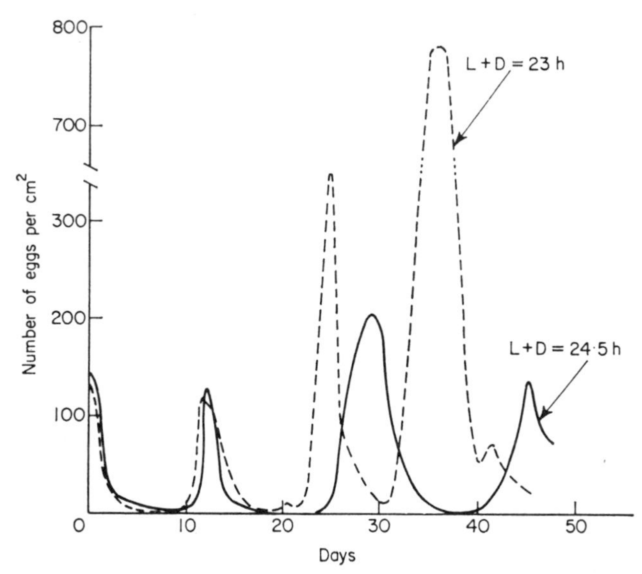

Fig. 5.4. The production of eggs by *Dictyota dichotoma* growing in the laboratory with light–dark cycles shorter than 24 h (13·5:9·5 = 23) (dashed line); and on light–dark cycles longer than 24 h (14·25:10·25 = 24·5 h) (solid line) (Vielhaben, 1963).

interval between egg production. Under LD 13.5 : 9.5, reproduction occurred at 11- to 12-day intervals, while under LD 14.25 : 10.25 the period of egg release was lengthened to 16–17 days. Vielhaben's controls were apparently reproducing every 14–15 days, rather than at 16-day intervals as had Müller's, but in her paper she gives no data for controls run at the same time as experimental schedules. Calculation shows that if the LD cycle is shortened to 23 h, then beats should occur at about 12 days as she observed, but long cycles like 24.5 h would be expected to produce beats only after 55 days!

There is no direct evidence in *Dictyota* for a tidal 12.4-h rhythm. The evidence for a component with a 24.8-h period is the shadow cast across the curve for photosynthesis as a function of time of day referred to previously (Vielhaben, 1963). The rate of photosynthesis, usually high during the day, falls in the afternoon earlier and earlier each day by 0.8 h. There is good evidence that *Dictyota* has a circadian rhythm in the time of gamete release that Vielhaben (1963) showed continued in LL.

In 1936, Hollenberg reported that *Halicystis*, the gametophytic stage of the coenocytic green filamentous alga *Derbesia*, which had been observed to form gametangia only at especially low "spring" tides twice each lunar month, could carry on this periodicity in the laboratory, removed from direct tidal influence. Extending this observation, Ziegler-Page and Kingsbury (1968) and Ziegler-Page and Sweeney (1968) made an extensive study of the reproduction of *Halicystis parvula* under carefully controlled conditions in the laboratory. They found that *Halicystis* did not show a semilunar periodicity in her experiments, either in a light–dark cycle or under LL. Reproduction however occurred quite regularly and in synchrony. The interval between the formation of gametangia was usually 4–5 days when growth conditions were optimal. This is a far cry from a semilunar periodicity. There was no effect of exposing the plants to light at night. Furthermore, keeping the algae in LD 8 : 8, a 16-h cycle, had no effect on the time of initiation of gametangia (Fig. 5.5). *Halicystis* was shown by Ziegler-Page and Sweeney (1968) to possess a well-defined circadian rhythm in the rate of photosyn-

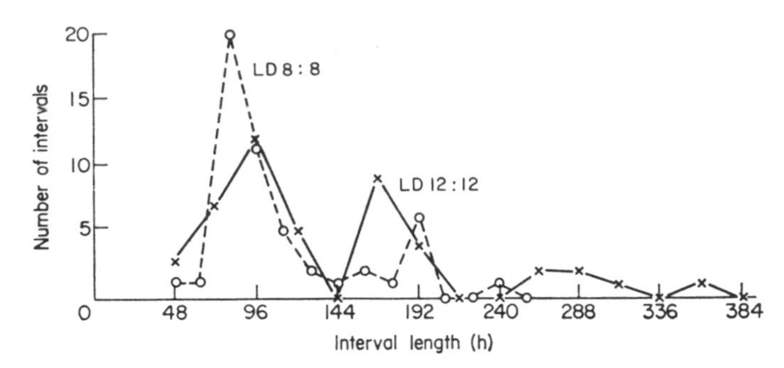

Fig. 5.5. The interval between successive gametangia in *Derbesia tenuissima* growing on LD 8 : 8, compared to that when plants are grown on LD 12 : 12. Changing the light–dark schedule did not drastically alter the timing of gametangial formation (Ziegler-Page and Sweeney, 1968).

thesis, persistent in LL. There was no evidence of any effect of a beat between this circadian and a tidal rhythm.

When *Halicystis* was maintained under lower than optimal light intensities, reproduction still occurred. The most interesting finding from these experiments was that under low light the interval between the formation of successive waves of gametangia on the same plant was either twice or three times the normal 4–5 days. Other intervals, such as 1.5 or 2.5 times the normal timing, were not observed (Fig. 5.6), as if a rhythm in gametangial formation with a 4- to 5-day period were still present and reproduction could only take place at a certain phase of this rhythm. Under unfavorable conditions for the formation of gametangia, some necessary prerequisite was not present when the first phase allowing reproduction occurred. Plants were forced to await the coming of a second, or perhaps even of a third, permissive phase to initiate gametangia. This phenomenon is quite similar to the phasing of cell division in dinoflagellates in the circadian cycle.

Marine animals and plants usually show tidal and semilunar, but not lunar, rhythms. However, true lunar cycles cued to the full moon have been observed in nature. One example is found in the Cahora Bassa Reservoir in Mozambique. Here several

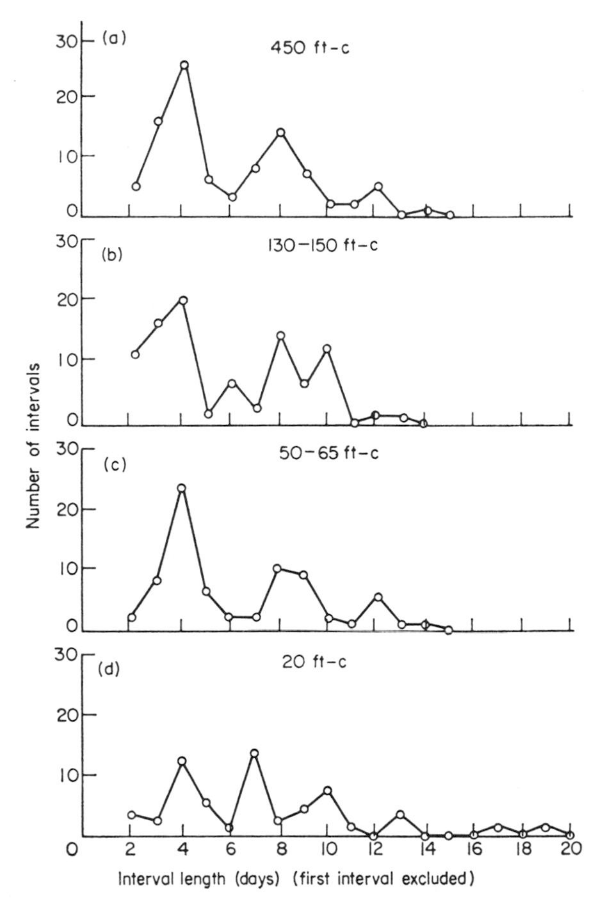

Fig. 5.6. The effect of light intensity on formation of gametangia. The intervals between successive gametangia in *Derbesia tenuissima* grown at different light intensities. Note that there are maxima at about 4-, 8- and 12-day intervals, but not at 5- and 6-day intervals (Ziegler-Page and Sweeney, 1968).

kinds of zooplankton are much more numerous just before full moon than at any other time. This periodicity is caused by the predation by a small fish that feeds on moonlight nights. During the dark of the moon, the zooplankton population can increase but during the feeding of this fish, *Limnothrissa miodon*, by moonlight, the population crashes (Fryer, 1986). What determines the intermittent feeding of the fish should be interesting to investigate.

Tidal and semilunar cycles are obviously of selective advantage to intertidal organisms and are common. They are not well understood, partly because many of these organisms are difficult to maintain for the necessary time in the laboratory. In addition, control of environmental conditions for such long times is difficult in practice. Study of these rhythms could prove very enlightening. Are they modified circadian rhythms or composites of more than one kind of periodicity? What environmental factors entrain them? What role does moonlight play in entrainment? The investigation of these questions and others are well worth the trouble to overcome the difficulties in studying these phenomena.

REFERENCES

Bohn, G. (1903). Sur les movements oscillatoires des *Convoluta roscoffensis. C. R. Seances Soc. Biol. Ses Fil.* **137**, 576–578.

Bracher, R. (1937). The light relations of *Euglena limosa* Card. Part I. The influence of intensity and quality of light on phototaxis. *J. Linn. Soc. London, Bot.* **51**, 23–42.

Bünning, E., and Müller, D. (1961). Wie messen Organismen lunare Zyklen? *Z. Naturforsch., B: Anorg. Chem., Org. Chem., Biochem., Biophys., Biol.* **16B**, 391–395.

Christie, A. O., and Evans, L. V. (1962). Periodicity in the liberation of gametes and zoospores of *Enteromorpha intestinalis* Link. *Nature (London)* **193**, 193–194.

Enright, J. T. (1963). The tidal rhythm of activity of a sand-beach amphipod. *Z. Vergl. Physiol.* **46**, 276–313.

Enright, J. T. (1976). Resetting a tidal clock: A phase-response curve for *Excirolana. Belle W. Baruch Libr. Mar. Sci.* **4**, 103–114.

Fauré-Fremiet, E. (1950). Rythme de marée d'une *Chromulina* psammophile. *Bull. Biol. Fr. Belg.* **84**, 207–214.

Fauré-Fremiet, E. (1951). The tidal rhythm of the diatom *Hantzschia amphioxys. Biol. Bull. (Woods Hole, Mass.)* **100**, 173–177.

Fauvel, P., and Bohn, G. (1907). Le rythme des marees chez les Diatomees littorales. *C. R. Soc. Biol.* **62**, 121–123.

Fryer, G. (1986). Lunar cycles in lake plankton. *Nature (London)* **322**, 306.

Hauenschild, C. (1960). Lunar periodicity. *Cold Spring Harbor Symp. Quant. Biol.* **25**, 491–497.

Hauenschild, C., Fischer, A., and Hoffmann, D. K. (1968). Untersuchungen am pazifischen Palolowurm *Eunice viridis* (Polychaeta) im Samoa. *Helg. Wiss. Meeresunters.* **18**, 254–295.

Hollenberg, G. J. (1936). A study of *Halicystis ovalis*. II. Periodicity in the formation of gametes. *Am. J. Bot.* **23**, 1–3.

Hopkins, J. T. (1966a). The role of water in the behavour of an estaurine mud-flat diatom. *J. Mar. Biol. Assoc. U. K.* **46**, 617–626.

Hopkins, J. T. (1966b). Some light-induced changes in behavour and cytology of an estaurine mud-flat diatom. *In* "Light as an Environmental Factor" (R. Bainbridge, G. C. Evans, and O. Rorkham, eds.), pp. 335–357. Wiley, New York.

Hoyt, W. D. (1927). The periodic fruiting of *Dictyota* and its relation to the environment. *Am. J. Bot.* **14**, 592–619.

Müller, D. (1962). Uber Jahres- und lunarperiodische Erscheinungen bei einiger Braunalgen. *Bot. Mar.* **4**, 140–155.

Neumann, D. (1976). Entrainment of a semilunar rhythm. *In* "Biological Rhythms in the Marine Environment" (P. J. DeCoursey, ed.), pp. 115–127. Univ. of South Carolina Press, Columbia.

Palmer, J. D. (1976). Clock-controlled vertical migration rhythms in intertidal organisms. *Belle W. Baruch Libr. Mar. Sci.* **4**, 239–255.

Palmer, J. D., and Round, F. E. (1965). Persistent, vertical migration rhythms in the benthic microflora. I. The effect of light and temperature on the rhythmic behavior of *Euglena obtusa*. *J. Mar. Biol. Assoc. U. K.* **45**, 567–582.

Palmer, J. D., and Round, F. E. (1967). Persistent vertical migration rhythms in the benthic microflora. VI. The tidal and diurnal nature of the rhythm in the diatom *Hantzschia virgata*. *Biol. Bull. (Woods Hole, Mass.)* **132**, 44–55.

Provasoli, L., Yamasu, T., and Manton, I. (1968). Experiments on the resynthesis of symbiosis in *Convoluta roscoffensis* with different flagellate cultures. *J. Mar. Biol. Assoc. U. K.* **48**, 465–479.

Rao, K. P. (1954). Tidal rhythmicity of rate of water propulsion in *Mytilus* and its modifiability by transplantation. *Biol. Bull. (Woods Hole, Mass.)* **106**, 353–359.

Round, F. E., and Happey, C. M. (1965). Persistent, vertical migration rhythms in benthic microflora. IV. A diurnal rhythm of the epipelic diatom association in non-tidal flowing water. *Bull. Br. Phycol. Soc.* **2**, 463–471.

Round, F. E., and Palmer, J. D. (1966). Persistent, vertical migration rhythms in benthic microflora. II. Field and laboratory studies on diatoms from the banks of the River Avon. *J. Mar. Biol. Assoc. U. K.* **46**, 191–214.

Smith, G. M. (1947). On the reproduction of some Pacific Coast species of *Ulva. Am. J. Bot.* **34**, 80–87.

Vielhaben, V. (1963). Zur Deutung des semilunaren Fortpflanzungszyklus von *Dictyota dichotoma. Z. Bot.* **51**, 156–173.

Walker, B. W. (1952). A guide to the grunion. *Calif. Fish and Game* **38**, 409–420.

Ziegler-Page, J., and Kingsbury, J. M. (1968). Culture studies on the marine alga *Halicystis parvula - Derbesia tenuissima.* II. Synchrony and periodicity in gamete formation and release. *Am. J. Bot.* **55**, 1–11.

Ziegler-Page, J., and Sweeney, B. M. (1968). Culture studies on the marine alga *Halicystis parvula - Derbesia tenuissima.* III. Control of gamete formation by an endogenous rhythm. *J. Phycol.* **4**, 253–260.

Rhythms That Match Environmental Periodicities: The Year

In all but the equatorial regions of the world, there is a change in the environment with season—a yearly cycle in temperature, light, and sometimes rainfall. These seasonal changes may be severe in temperate latitudes. Plants and animals have evolved a whole series of adaptive changes that allow them to survive through the inclement part of the year. Plants form seeds or tubers, while vertebrates hibernate and insects go into diapause. On the other hand, growth and reproduction take place during the part of the year when mild weather will favor the survival of offspring.

The key to successful adaptation to a recurrent pattern of change is preparedness. Thus, as one would expect, it is not the coming of the conditions for which the adaptations are designed that sets off their initiation. A seed that is not formed until the snow flies might never be formed at all. The question that interests us here is, what is the nature of the devices that make possible the detection of the coming of winter or the coming of spring before either actually arrives. There are at least two possibilities that come to mind. First, organisms might contain mechanisms that are able to measure time in years, a rhythm with an annual period, just as circadian rhythms may be used to measure days. Second, organisms might make use of the fact that the day length changes with the seasons, shorter as winter approaches and longer with the coming of spring. They would then only require a device for measuring the length of the day with respect to the night. It seems reasonable that one might find both of these methods in use, were one to canvass enough different organisms in one's search. It has been found, however, that the second method of predicting the coming season, the measurement of the length of the day, appears to be far the most common, both in plants and animals. I shall discuss this kind of yearly periodicity, photoperiodism, first and then return to a consideration of the evidence for the existence of true annual rhythms.

THE MEASUREMENT OF DAY LENGTH IN PHOTOPERIODISM

The importance of the length of day in initiating the change from the vegetative to the reproductive phase of growth was first discovered by Garner and Allard (1920) in tobacco and soy bean, plants that they showed initiated flower buds only when the days were *shorter than* a certain number of hours. Other plants were found to flower only when the days were *longer than* a given duration. Thus, there are "short-day" plants and "long-day" plants. Even some red algae respond to day length with changes in morphology (Iwasaki, 1961). The critical day length where the developmental pattern changes is about 12 h in both types, although it varies somewhat from species to species. Seasonal changes in animals, such as change in hair color before winter and growth of gonads and reproductive behavior before spring, were later found to also be cued by day length in much the same way as in plants. The question of how all these organisms measure the length of day or night is both interesting and important.

Two kinds of theories have evolved to account for photoperiodic time measurement and the proponents of each defended their views with vigor if not with rancor. Just as time may be measured by using an hourglass or an electric clock, so can the length of day be measured by the accumulation or disappearance of some cellular component (the sand of the hourglass), or by a comparison of the external light or darkness with an internal clock counting hours, a circadian clock in every way like those described in Chapter 3.

Proponents of the "hourglass" theory for the time measurement in photoperiodism hunted for a component that appeared or disappeared in light or in darkness with the appropriate rate to provide the necessary yardstick. It was soon apparent that, in short-day plants at least, the length of the night rather than the day length was the parameter being measured. This conclusion was reached as a result of experiments in which the night was interrupted by a very brief light exposure, only a few minutes. In short-day plants exposed to short days and hence long nights, such an interruption completely blocked flowering, although the control plants given the same light–dark schedule but without the light interruption, flowered as expected. A similar situation was discovered in long-day plants as well, where however the long night prevented flower formation in the controls as expected and the light interruption now invoked flowering. Thus, short days and long nights interrupted by a very short light exposure were comparable in every way to long days and short nights. That exposure to very dim light at night could so strongly affect the flowering of photoperiodic plants led Bünning (1971; Bünning and Moser, 1969) to suggest that moonlight might upset photoperiodic induction and that the selective advantage of circadian leaf movements might be in reducing exposure to moonlight!

Action spectra for the effect of light as an interruption in a long night (Borthwick *et al.,* 1952) showed that phytochrome (the P_r form) was the photoreceptor and that the far-red-absorbing P_{fr} was active. Since phytochrome can be converted from one form to the other by light, it was suggested (Hendricks, 1960, 1963) that the substance timing the length of the night was phytochrome in the P_r form. At the end of a day, most of the phytochrome is found in the P_{fr} form. During the night this form disappears and P_r appears by synthesis or reversal from P_{fr}. This is a slow process, but not really slow enough to time a whole night. In the cotyledons of corn and other monocots, reappearance of P_r is complete in about 4 h (Butler *et al.*, 1963), while in etiolated seedlings of the dicots pea and *Sinapsis,* the P_{fr} disappeared even faster in darkness. The rate of phytochrome conversion was measured in both long-day and short-day plants by Hopkins and Hillman (1965) who also found that no further change occurred after 4–5 h in darkness. Therefore, why should the critical night length be 12 h, not 4 h? Furthermore, it has been shown (Salisbury, 1963) that the critical night length does not vary much as the temperature is changed. The conversion of phytochrome is much more temperature dependent.

What evidence is there that the circadian clock is the timer for the measurement of the length of the night in photoperiodism? The temperature independence of the critical night length suggests a circadian timer. Bünning (1936) was the first to propose that the same timing mechanism was responsible for photoperiodic time measurement and the diurnal rhythms he was studying in the sleep movements of bean leaves. He distinguished two alternating phases of the circadian cycle which he called "photophil," or light-loving, and "scotophil," dark-loving. He postulated that light falling on the plant in the scotophil phase would inhibit flowering, and in the photophil phase, enhance flower production. The change from one state to the other would occur about every 12 h, hence accounting for the 12-h critical day length in both short- and long-day plants. While short-day plants fit nicely into this scheme, further assumptions are needed to account for the existence of long-day plants. Bünning postulated that in these plants the photophil phase lasts longer than 12 h, thus extending into the first part of the night except during the long days of summer.

Bünning reasoned that if the circadian clock were really responsible for measurement of day length, leaf position in a plant possessing leaf sleep movements should indicate whether the plant was in the photophil (or day phase) or scotophil (or night phase) at any given time (Bünning, 1960). It is my impression that only some photoperiodic plants show this correlation. Other types of experiments have shown that Bünning was essentially correct about the importance of circadian timing. These experiments were designed to examine the effects of light interruptions during a night longer than 12 h. One of the first of these was done by Claes and Lang (1947) using the long-day plant *Hyoscyamus* grown with a 48-h night. Flowering was promoted as expected by light interruptions early in the dark period. The effectiveness increased to a maximum as the light interruption was placed later in the night, then flowering declined as still later light interruptions were assayed, and the night became too long to allow flowering. But, the surprise, later still the light again *increased* flowering. Claes and Lang explained their results by suggesting that the light breaks in the early evening were interpreted by the plant as extensions of the previous day, while the very late light breaks were seen as part of the following day and under both circumstances the long day promoted flowering. However, their findings could also be interpreted as evidence for a circadian rhythm in sensitivity to a light interruption. To decide between these two alternatives, light interruptions in dark periods longer than 48 h were investigated. Such experiments clearly showed a circadian rhythm in the sensitivity to light interruptions, both in short-day plants such as *Kalanchoe* (Carr, 1952; Bünsow, 1960; Engelmann, 1960) and *Chenopodium rubrum* (Cumming *et al.*, 1965) and in long-day plants like *Sinapsis alba* (Kinet *et al.*, 1973). Light during the night phase inhibited flowering in short-day plants and promoted it in long-day plants, even when given during a 60-h dark period (Fig. 6.1). Maxima were observed at about 22-h intervals, three such peaks in long-day plants and two in short-day plants. In *Kalanchoe blossfeldiana*, not only did peaks in the induction of flowering occur at 22- to 23-h intervals when a 62-h dark period was interrupted with red light at different times, but these peaks corresponded with the midday maximum in the circadian rhythm in petal movement (Fig. 6.2). A similar rhythmic flowering was induced in soy bean plants kept for seven cycles in schedules of 8 h light alternating with 64 h of darkness interrupted by light pulses at different times (Fig. 6.3).

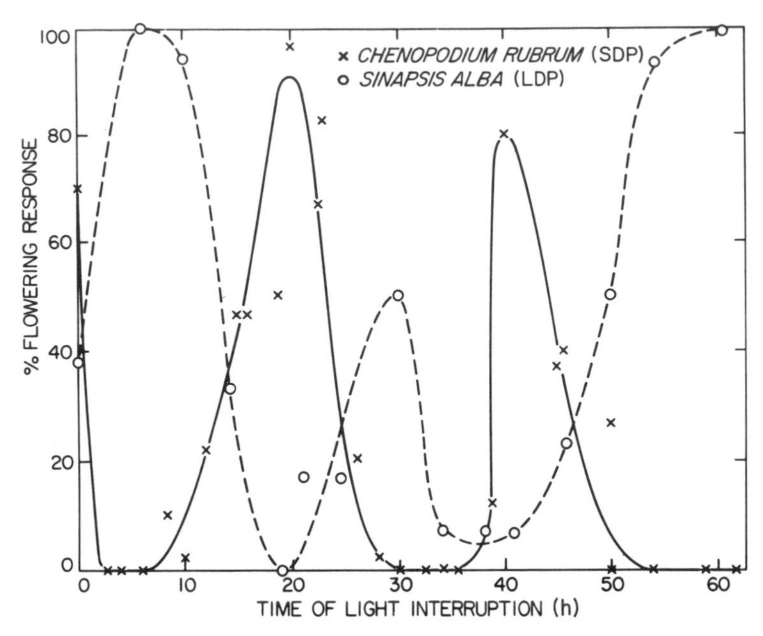

Fig. 6.1. Rhythmic sensitivity of flower induction to light interruptions during a 60-h-long night. Solid line and crosses, data from *Chenopodium rubrum,* a short-day plant, interruptions, 4 min red light. Broken line and circles, data from *Sinapsis alba*, a long-day plant, light interruptions 8 h white fluorescent light. [Fig. 1 in Sweeney, 1973; data redrawn from Cumming *et al.* (1965) and Kinet *et al.* (1973).]

A marked periodicity is apparent in this and other similar experiments, suggesting that the time of the interrupting light pulse is sensed by a circadian mechanism. The period of circadian rhythms is characteristically insensitive to the ambient temperature. A similar temperature independence was found for the interval between maxima in flowering in the short-day plants, *Xanthium pennsylvanicum* and *Pharbitis nil*, although the number of flowers increased at higher temperatures (Fig. 6.4). This finding strengthens the hypothesis that the circadian clock is responsible for the measurement of night length.

To show the presence of a circadian rhythm in photoperiodic responsiveness, it is not necessary to rely on light interruptions of long dark periods. Superimposable rhythmic data for the flowering response of *Chenopodium rubrum* were obtained from an experiment where the length of the dark period was varied from 3 to 96 h in one set of plants and light interruptions at different times were given to another set at different times in a night 72 h long (Fig. 6.5).

A potential problem for the interpretation of the role of circadian rhythmicity in experiments with light breaks or with repeated cycles of very long dark periods arises because circadian rhythms are known to be reset by exposure to light. Most plants require a number of days of the proper length to flower. For example, in order to flower, soybeans require at least seven cycles of long nights. If the rhythmicity were phase-shifted by the light breaks at night, when rhythms are in fact most sensitive to phase-shifting by light, then the phase of a circadian rhythm would become hopelessly confused. It seems that the light breaks are either too short or too dim to reset or, alternatively, the circadian rhythm is reset each day by the main light period. In

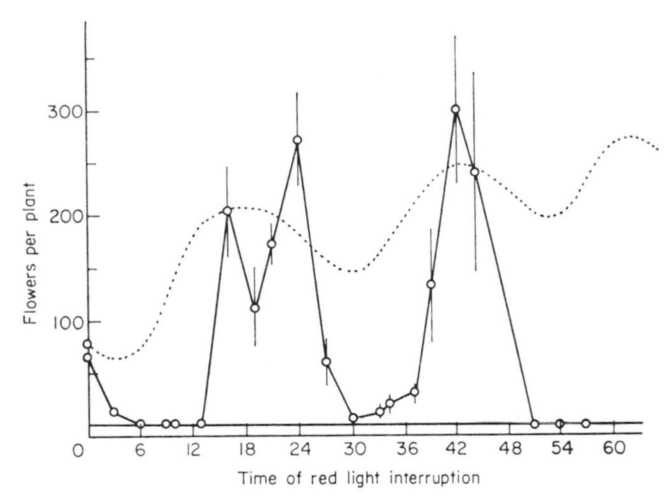

Fig. 6.2. The number of flowers induced in *Kalanchoē* by a 62-h dark period interrupted with red light at different times. Lines through the points represent uncertainty. The dotted curve is the course of petal movement during a similar long dark period (after Engelmann, 1960).

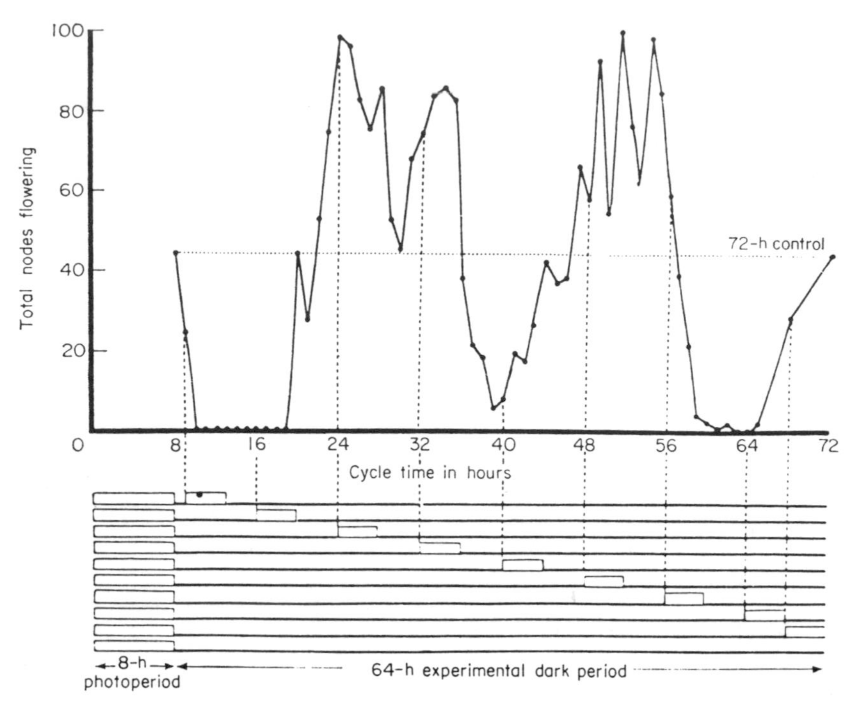

Fig. 6.3. Flowering of soya bean following 7 cycles of 7 h light, 64 h of darkness interrupted at various times with a 4-h interruption with bright white light. The abscissa is the time in the dark period when the light interruption began. The light schedule is diagramed below (after Coulter and Hamner, 1964).

light interruption experiments, red is clearly the effective wavelength and phytochrome is the photoreceptor in plants. In *Phaseolus*, resetting of the leaf sleep movements was shown by Bünning and Moser (1966) to be brought about by red light, but the exposure required was much longer, 3.5 h at least, and at a higher intensity than required for an effective light interruption. No true far-red photoreversal of resetting could be observed in their experiments either. The experiment with *P. nil* graphed in Fig. 6.4 clearly shows the lack of resetting by 5-min exposure to red light, since the first red light pulses near 8 h of darkness had no effect on the position of the second peak in flowering.

There is thus very strong evidence for the involvement of a circadian rhythm in flower induction. But what role does this rhythm play? Is it the clock by which the length of the day and night are measured in photoperiodic organisms? Bünning was right that light coinciding with night, the scotophil phase, changes the flowering response in both long- and short-day plants. This is true even when the scotophil phase occurs in a circadian rhythm in an otherwise long dark period. The coincidence of light with a part of the scotophil phase appears to be necessary for the measurement of day length.

A fortuitously designed experiment of Hillman's with *L. perpusilla* provides the best evidence that timing of day length is inseparably connected with the phase of a circadian rhythm (Hillman, 1964). *Lemna* can be grown in bacteria-free culture and fed with glucose so that it can be reared independently of requirements for light for photosynthesis. Thus, long light periods can be eliminated and "skeleton" photoperiods, where short pulses of light divide darkness into segments, can be used in experiments on the induction of flowering. In such experiments, Hillman found that *Lemna* will flower when given only two 15-min light exposures in every 24 h, if the light and dark were

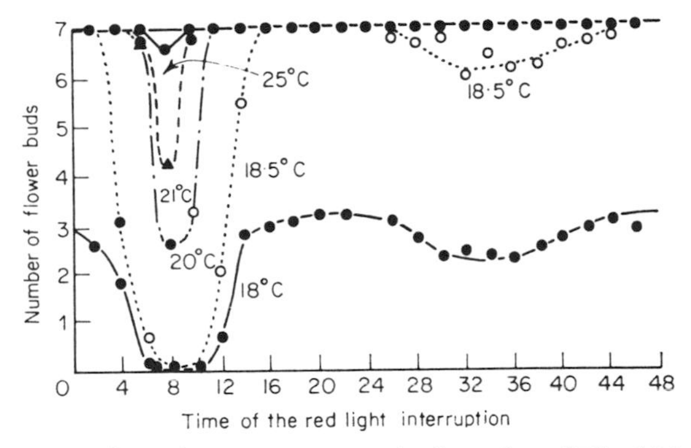

Fig. 6.4. The effect of temperature on the flowering of *Pharbitis nil* following a single dark period interrupted at different times with 5 min of red light, as shown on the abscissa. The periodicity is clearly independent of temperature although the number of flowers is increased by raising the temperature (after Takimoto and Hamner, 1964).

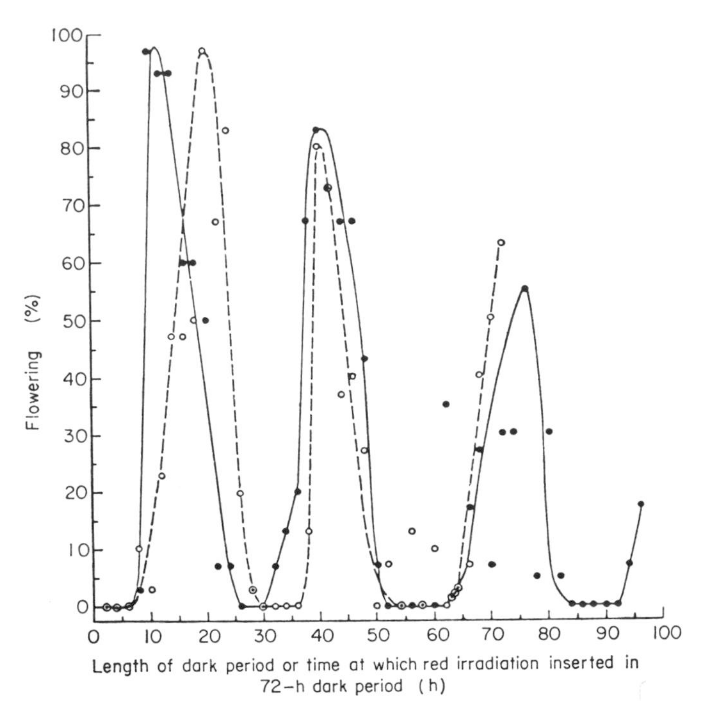

Fig. 6.5. The flowering of *Chenopodium rubrum* following a single dark period of different lengths from 3 to 96 h. (●) or a single 72-h dark period interrupted at various times by 4 min of red light (○). Plants were otherwise in continuous light (after Cumming, *et al.*, 1965).

arranged in the sequence 15 min light, 13 h darkness, 15 min light, 10.5 h darkness. Further experiments showed that the sequence of these light and dark treatments is of the utmost importance. When the *Lemna* is transferred from the continuous light under which the stock cultures are grown, the first dark period must be the *long* one for flowering to occur, that is, the schedule must be: 13 D:0.25 L:10.5 D:0.25 L and not 10.5 D:0.25 L:13 D:0.25 L. It seems reasonable to conclude that the short light periods mark the beginning and the end of two qualitatively different dark periods, one interpreted by the plant as day and the other as night. But can we tell which is which? If flowering occurs in this short-day plant, the 13-h dark period must have been interpreted as night. Therefore, the first dark period and every alternate dark period thereafter is considered a night. This explains why no flowering occurred on the schedule that began with the shorter dark period, even though the second dark period was longer than the critical night. But why is the first dark period always interpreted as a night by *Lemna*? The behavior of circadian rhythms explains this easily: like other organisms in long-continued bright light, the circadian rhythmicity in *Lemna* has damped out preceding the first dark period. The first transfer to darkness reinitiates rhythmicity, and as always the rhythm starts with the beginning of the night phase, CT 12. A 13-h dark period will only act as a long night for flower induction when it coincides with this night phase, never the day phase.

Hillman extended his experiments to confirm this interpretation by varying only the length of the first dark period after continuous light, and then giving a schedule of either 0.25 L: 10.5 D: 0.25 L: 13 D. or 0.25 L: 13 D: 0.25 L: 10.5 D. As in the first experiments, promotion of flowering was altogether determined by the length of the *first* dark period. With the first schedule, the most flowers were induced when the first dark period, the variable one, was 15 or 39 h long. With the second light–dark schedule, most effective flower induction occurred following a preliminary dark period of 24 or 48 h (Fig. 6.6), just the reverse of the results with the first schedule. These results would be very difficult to explain without invoking a circadian rhythm controlling time measurement, but they fall into place naturally when circadian lore is taken into account, as follows: as in Hillman's first experiment, the *Lemna* stock cultures that have been kept in constant light for a long time are arrhythmic. The rhythm is started by the first dark period; that is why it is the determining one. The phase of the rhythm is also set by the beginning of this dark period at 12 CT, the beginning of the night phase. The 15-min light exposures in the skeleton photoperiod are not sufficiently bright or long to change the phase of the circadian rhythm, which continues as initiated at the beginning of the first dark period. The behavior of the circadian rhythm in

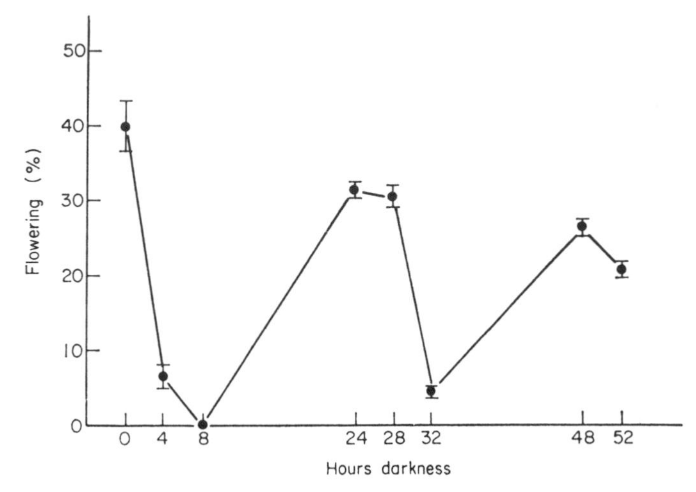

Fig. 6.6. The flowering of *Lemna perpusilla* treated with a dark period of different lengths preceding seven repetitions of the schedule ¼L : 13D : ¼L : 10½D. Twice the standard error of each mean is shown by brackets around each point (after Hillman, 1964).

CO_2 output has been followed in *Lemna* where CO_2 output can be monitored continuously unlike flowering (Hillman, 1971): the results fully corroborate the earlier conclusions (Hillman, 1964). *Lemna* flowers whenever a dark period longer than the critical night *coincides with the night phase* of its circadian clock. Darkness during the day is ineffective in initiating flowering. Therefore it seems reasonable that the circadian clock provides the time base to which the actual light and darkness of the environment are compared (Pittendrigh, 1966). The mechanism of comparison is still unknown, another effect of light that does not change the phase of the circadian clock, perhaps an effect mediated through phytochrome (Heide, 1977).

It is interesting to note that short photoperiods of the skeleton type, if properly arranged with respect to circadian time, will induce photoperiodic effects in insects (Minis, 1965; Pittendrigh and Mimis, 1971; Atkisson, 1966) and in birds (Menaker and Eskin, 1967). In these organisms, too, there is clear evidence that two kinds of light effects can change the photoperiodic responses: one kind of light effect acts to determine the phase of the circadian clock and the other senses the coincidence between environmental light or darkness and the phase of the circadian clock.

Since the circadian clock is clearly implicated in photoperiodic time measurement, the design of future experiments on photoperiodism must take careful consideration of the phase relationships of circadian systems and the manner in which they are reset. Experiments in which the phase or the period of the circadian clock is changed by drug treatment and the subsequent behavior of the critical night length is examined could enlighten us concerning *how* clock-time and the critical night are connected. As far as I know, such experiments have not yet been carried out.

EVIDENCE FOR THE EXISTENCE OF TRUE CIRCANNIAN RHYTHMS IN PLANTS

A rigorous demonstration of a true endogenous annual or "circannian" rhythm in flowering plants has probably never been achieved. The difficulties standing in the way of detecting such a rhythm are formidable. First, environmental conditions of light and temperature must be kept strictly constant for at least three cycles, i.e., 3 years, not an easy task since equipment is subject to failure. Second, annual variations in the composition of media must be avoided. This poses an almost insurmountable difficulty, since even twice-distilled water has been shown to vary in organic composition, depending on whether or not the reservoir source of the water before purification contains algae or other organisms in bloom proportions. These blooms themselves tend to be seasonal. Photoperiodic control must be ruled out when considering whether or not periodicities that appear to be annual are really annual rhythms in the strict sense. Even tropical plants that grow in an environment with a minimum variation in day length with season may be controlled photoperiodically. For example, the dropping of the leaves of the tropical shrub *Plumaria* have been shown to be photoperiodic (Murashige, 1966). Tropical forest trees do sometimes show cycles of defoliation where different individual trees in the same locality or even different branches of the same tree lose their leaves at different times. A response to day length can hardly explain such behavior. However, trees are intractible laboratory organisms, difficult to investigate under controlled laboratory conditions.

Small plants grown under rigorously controlled temperature and unchanging light conditions have been observed to vary in growth rate over the year. Two studies by Pirson and Göllner (1953) and Bornkamm (1966) have detected annual periodicity in *Lemna minor* L. under constant conditions. Pirson and Göllner measured the growth of *Lemna* roots, using as an indication the number of hours required for the roots to increase in length from 1 to 20 mm. In December, this time was twice as long as in the preceding or following June. The decrease in the rate of root growth was closely correlated with increase in osmotic value of the root cells and decrease in the time required to plasmolyse these cells. In Bornkamm's experiments, the growth of *Lemna*, measured this time by increase in dry weight, was greater in summer although the plants were kept in constant light (1700 lux) and the temperature was maintained all year at 25 °C. This was observed in 2 successive years. Increased rate of dry weight accumulation was accompanied by an increase in protein content and in the ratio of protein to carbohydrate. In the third year of Bornkamm's observations, however, no summer maximum was observed.

Annual variation in the growth rate of plant tissue cultures has also been noticed. For example, S. Lavee (personal communication) has observed that the cells of both dwarf and tall apple and of olive trees in culture show annual cycles of growth, measured either as wet or dry weight. Cultures were kept at 27–28 °C under constant light. This yearly cycle appeared in 5 consecutive years, although the amplitude became somewhat reduced with time. Both this and the work with *Lemna* depend on growth on artificial media and hence are subject to error arising from undetected changes in the composition of the medium as mentioned above. In a recent study where an effort to use an entirely artificial medium always made up with identical water, Yentsch *et al.* (1980) measured an annual rhythm in growth rate of the marine dinoflagellate *Gonyaulax tamarensis*. This study is still in progress.

If there are annual rhythms in growth, it is not surprising that annual differences in the biochemistry of cells can also be detected. Kessler and Czygan (1963) made the interesting observation that in intact cells of the green alga *Ankistrodesmus braunii*, a seasonal difference in the capacity to reduce nitrate could be demonstrated in extracts prepared at different times of year, maxima being found in June and July and minima in October and November in 2 successive years, although the cells were grown in constant light and temperature. Using *Chlorella pyrenoidosa*, Kessler and Langner (1962) showed a similar annual variation in hydrogenase activity. Maxima were observed again in the summer, the rate being five times that at the minimum in March. However, the cultures were grown in a window and hence were surely subject to annual fluctuations in light intensity, temperature, and day length.

It is obviously easier to maintain a dormant organism under strictly constant conditions for the long time required to demonstrate an annual rhythm than to do so with growing plants or animals. Seeds, kept air-dried and tested for germination from time to time, sometimes show annual rhythms in the capacity to germinate. Bünning and Müssle (1951) and Bünning and Bauer (1952) reported annual rhythms from such studies of the seeds of *Digitalis lutea, Chrysanthemum corymbosum*, and *Gratiola officinalis*. The seeds of *Digitalis* and *Chrysanthemum* were stored at widely different temperatures, −22, 4, 45, and 55 °C, and then the germination of samples was tested at 25 °C. The storage temperature did not much affect the germination capacity, in which maxima were observed in early spring and minima in late summer. The gaseous atmosphere during storage (N_2, O_2, or CO_2) did not greatly affect the germination of *Gratiola*, which germinated readily in January, February, and March but not at all during the spring and summer. In no case were the seeds examined for a long enough time to observe more than two cycles.

A recent example of a study with long storage of dormant material concerns the cysts of a marine dinoflagellate. The dinoflagellate *Gonyaulax tamarensis*, from the northeast coast of the United States, encysts under conditions unfavorable for growth in winter. The cysts germinate and give rise to motile cells when brought back to favorable conditions for growth, such as new medium and a higher temperature. Anderson and Keafer (1987) have been studying the germination of cysts from the vicinity of Woods Hole, both immediately after collection from nature and after long storage in the laboratory in the cold. They find that germination occurs with a definite yearly rhythm, of which they have now measured two cycles.

Annual rhythms are better known in animals than in plants. Several mammalian species, including the golden-mantled ground squirrel (Pengelley and Kelly, 1966; Pengelley and Asmundson, 1969) and the woodchuck (Davis, 1967), have clear annual rhythms in fat accumulation and hibernation, even under laboratory conditions of constant LD or LL and constant temperature. The migratory restlessness in warblers seems also to have an annual rhythmic component (Gwinner, 1971). A cave-dwelling crayfish has also been reported to show a circannian rhythm (Jegla and Poulson, 1970).

Both photoperiodic responses to night length and endogenous annual rhythms have great survival value in a seasonally changing environment. It is not surprising that they are encountered so commonly in all kinds of organisms.

REFERENCES

Anderson, D. H., and Keafer, B. A. (1987). An endogenous annual clock in the toxic marine dinoflagellate *Gonyaulax tamarensis*. *Nature (London)* **325**, 616–617.

Atkisson, P. L. (1966). Internal clocks and insect diapause. *Science* **154**, 234–241.

Bornkamm, R. (1966). Ein Jahresrhythmus des Washstums bei *Lemna minor* L. *Planta* **69**, 178–186.

Borthwick, H. A., Hendricks, S. B., Parker, M. W., Toole, E. H., and Toole, V. K. (1952). A reversible photoreaction controlling seed germination. *Proc. Natl. Acad. Sci. U. S. A.* **38**, 662–666.

Bünning, E. (1936). Die endogene Tagesrhythmik als Grundlage der photoperiodischen Reaktion. *Ber. Dtsch. Bot. Ges.* **54**, 590–607.

Bünning, E. (1960). Circadian rhythms and the time measurement in photoperiodism. *Cold Spring Harbor Symp. Quant. Biol.* **25**, 249–256.

Bünning, E. (1971). The adaptive value of leaf circadian movements. *In* "Biochronometry" (M. Menaker, ed.), pp. 203–211. Natl. Acad. Sci., Washington, D. C.

Bünning, E., and Bauer, E. W. (1952). Uber die Ursachen endogener Keimfahigkeitsschwabkungen in Samen. *Z. Bot.* **40**, 67–76.

Bünning, E., and Moser, I. (1966). Response-Kurven bei der circadainen Rhythmik von *Phaseolus*. *Planta* **69**, 101–110.

Bünning, E., and Moser, I. (1969). Interference of moonlight with the photoperiodic measurement of time by plants, and their adaptive reaction. *Proc. Natl. Acad. Sci. U. S. A.* **62**, 1018–1022.

Bünning, E., and Müssel, L. (1951). Der Verlauf der endogenen Jahresrhythmik in Samen unter dem Einfluss verschiedenartiger Aussenfaktoren. *Z. Naturforsch., B: Anorg. Chem., Org. Chem., Biochem., Biophys., Biol.* **6B**, 108–112.

Bünsow, R. C. (1960). The circadian rhythm of photoperiodic responsiveness in *Kalanchoe*. *Cold Spring Harbor Symp. Quant. Biol.* **25**, 257–260.

Butler, W. L., Lane, H. C., and Siegelman, H. W. (1963). Nonphotochemical transformations of phytochrome *in vivo*. *Plant Physiol.* **38**, 514–519.

Carr, D. J. (1952). A critical experiment on Bunning's theory of photoperiodism. *Z. Naturforsch., B: Anorg. Chem., Org. Chem., Biochem., Biophys., Biol.* **7B**, 570–571.

Claes, H., and Lang, A. (1947). Die Blutenbildung von *Hyoscyamus niger* in 48-stundigen Licht-Dunkel-Zyklen und in Zyklen mit aufgeteilten Lichtphasen. *Z. Naturforsch., B: Anorg. Chem., Org. Chem., Biochem., Biophys., Biol.* **2B**, 56–63.

Coulter, M. W., and Hamner, K. C. (1964). Photoperiodic flowering response of Biloxi soybean in 72-hour cycles. *Plant Physiol.* **39**, 848–856.

Cumming, B. C., Hendricks, S. B., and Borthwick, H. A., (1965). Rhythmic flowering responses and phytochrome changes in a selection of *Chenopodium rubrum*. *Can. J. Bot.* **43**, 825–853.

Davis, D. E. (1967). The annual rhythm of fat deposition in woodchucks *(Marmota monax)*. *Physiol. Zool.* **40**, 391–402.

Engelmann, W. (1960). Endogene Rhythmik und photoperiodische Bluhinduktion bei *Kalanchoe*. *Planta* **55**, 496–511.

Garner, W. W., and Allard, H. A. (1920). Effect of relative length of day and night and other factors of the environment on the growth and reproduction in plants. *J. Agric. Res.* **18**, 553–606.

Gwinner, E. (1971). A comparative study of cirannual rhythms in warblers. *In* "Biochronometry" (M. Menaker, ed.), pp. 405–427. Natl. Acad. Sci., Washington, D. C.

Heide, O. M. (1977). Photoperiodism in higher plants: An interaction of phytochrome and circadian rhythms. *Physiol. Plant.* **39**, 25–32.

Hendricks, S. B. (1960). Rates of change of phytochrome as an essential factor determining photoperiodism in plants. *Cold Spring Harbor Symp. Quant. Biol.* **25**, 245–248.

Hendricks, S. B. (1963). Metabolic control of timing. *Science* **141**, 21–27.

Hillman, W. S. (1964). Endogenous circadian rhythms and the response of *Lemna perpusilla* to skeleton photoperiods. *Am. Nat.* **98**, 323–328.

Hillman, W. S. (1971). Carbon dioxide output as an index of circadian timing in *Lemna* photoperiodism. *In* "Biochronometry" (M. Menaker, ed.), pp. 251–271. Natl. Acad. Sci., Washington, D. C.

Hopkins, W. G., and Hillman, W. S. (1965). Phytochrome changes in tissues of dark-grown seedlings representing various photoperiodic classes. *Am. J. Bot.* **52**, 427–432.

Iwasaki, H. (1961). The life cycle of *Porphyra tenera in vitro. Biol. Bull. (Woods Hole, Mass.)* **121**, 173–177.

Jegla, T. C., and Poulson, T. L. (1970). Circannian rhythms. I. Reproduction in the cave crayfish, *Orconectes pellucidus inermis. Comp. Biochem. Physiol.* **33**, 347–355.

Kessler, E., and Czygan, C. F. (1963). Seasonal changes in the nitrate-reducing activity of a green alga. *Experientia* **19**, 89–90.

Kessler, E., and Langner, W. (1962). Jahresperiodische Aktivitatsschwankungen bei einer *Chlorella. Naturwissenschaften* **49**, 331–382.

Kinet, J. M., Bernier, J. M., Bodson, M., and Jacqmard, A. (1973). Circadian rhythms and the induction of flowering in *Synapsis alba. Plant Physiol.* **51**, 598–600.

Menaker, M., and Eskin, A. (1967). Circadian clock in photoperiodic time measurement: A test of the Bunning hypothesis. *Science* **157**, 1182–1185.

Minis, D. H. (1965). Parallel peculiarities in the entrainment of a circadian rhythm and photoperiodic induction in the pink boll worm *(Pectinophora gossypiella). In* "Circadian Clocks" (J. Aschoff, ed.), pp. 333–343. North-Holland Publ., Amsterdam.

Murashige, T. (1966). The deciduous behavior of a tropical plant, *Plumaria acuminata. Plant Physiol.* **19**, 348–355.

Pengelley, E. T., and Asmundson, S. M. (1969). Free-running periods of endogenous circannian rhythms in the golden-mantled ground squirrel, *Citellus lateralis. Comp. Biochem. Physiol.* **30**, 177–183.

Pengelley, E. T., and Kelley, K. H. (1966). A "circannian" rhythm in hybernating species of the genus *Citellus* with observations on their physiological evolution. *Comp. Biochem. Physiol.* **19**, 603–617.

Pirson, A., and Göllner, E. (1953). Beobachtungen zur Entwicklungsphysiologie der *Lemna minor* L. *Flora (Jena)* **139**, 485–498.

Pittendrigh, C. S. (1966). The circadian oscillation in *Drosophila pseudoobscura* pupae: A model of the photoperiodic clock. *Z. Pflanzenphysiol.* **54**, 275–307.

Pittendrigh, C. S., and Minis, D. H. (1971). The photoperiodic time measurement in *Pectinophora gossypiella* and its relation to the circadian system of that species. *In* "Biochronometry" (M. Menaker, ed.), pp. 212–250. Natl. Acad. Sci., Washington, D. C.

Salisbury, F. B. (1963). Biological timing and hormone synthesis in flowering of *Xanthium. Planta* **59**, 518–534.

Sweeney, B. M. (1973). The temporal regulation of morphogenesis in plants, hourglass and oscillator. *Brookhaven Symp. Biol.* **25**, 95–110.

Takimoto, A., and Hamner, K. C. (1964). Effect of temperature and preconditioning on photoperiodic response of *Pharbitis nil. Plant Physiol.* **39**, 1024–1030.

Yentsch, C. M., Mague, F., Glover, H., and Yaskowski, K. (1980). Evidence of an apparent annual rhythm in the toxic red tide dinoflagellate *Gonyaulax tamarensis. Int. J. Chronobiol.* **7**, 77–84.

7

Rhythms That Do Not Match Environmental Periodicities

Besides the rhythms with periods that match the common oscillations of the environment, there is a whole array of phenomena that repeat at intervals of every imaginable length. In some the period is very short, as in the beating of flagella or cilia, too fast for the eye to follow. In some it is so long that it is barely detectable in a human lifetime. The rhythms about which we are speaking here cannot be cued by the environment because there are no environmental variables with matching frequencies. Of course these rhythms are interesting in themselves, but they have an added attraction to students of circadian rhythmicity because they may represent the raw material upon which evolution has acted to select rhythms with periods that do match environmental periodicities such as the day. It is outside the scope of this book and the ability of this author to describe every one of this multitude. Only a few examples have been selected to give the reader an idea of their diversity.

RHYTHMS WITH SHORT PERIODS OF SECONDS TO MINUTES

The beating flagellum of a sperm of the sea squirt *Ciona* has been photographed with a light flashing at a rate of 200/sec (Yates, 1986). These photographs show that the period of the undular motion of the flagellum is about $\frac{1}{40}$ of a second. In addition, a wave of contraction passes over the whole surface of organisms that are covered with many cilia, so that the timing of the beat of individual cilia must be coordinated by a secondary rhythmicity. The rate at which the flagella beat is markedly temperature dependent, at least in organisms where this has been investigated. Thus, temperature-compensating mechanisms do not seem to have evolved in cilia and flagella, nor is any selective advantage of such a mechanism immediately obvious. In general, temperature dependence does not seem disadvantageous in a rhythmicity that does not serve a time-keeping function.

STREAMING IN A SLIME MOLD

Another rhythmic movement has been measured most ingeniously by Kamiya. This is the alternation in the direction of flow of the plasmodium of the slide mold *Physarum polycephalum*. Plasmodia were persuaded to flow back and forth between two small chambers via a pore. The pressure within one of these chambers could be changed continuously so that the force was always just enough to prevent movement of the plasmodium. Beautiful records of the pressure applied in this way were obtained (Fig. 7.1). Furthermore, substances could be introduced into either chamber and their effect on streaming quantified. When both chambers were filled with N_2, the amplitude of the oscillation increased; inspection of Fig. 7.1 shows that the period, about 3 min before anaerobiosis, was slowed to approximately 4 min by the presence of nitrogen. Adding 10^{-3} M KCN produced a change in both amplitude and frequency similar to that caused by N_2. Dinitrophenol appeared to have quite a dif-

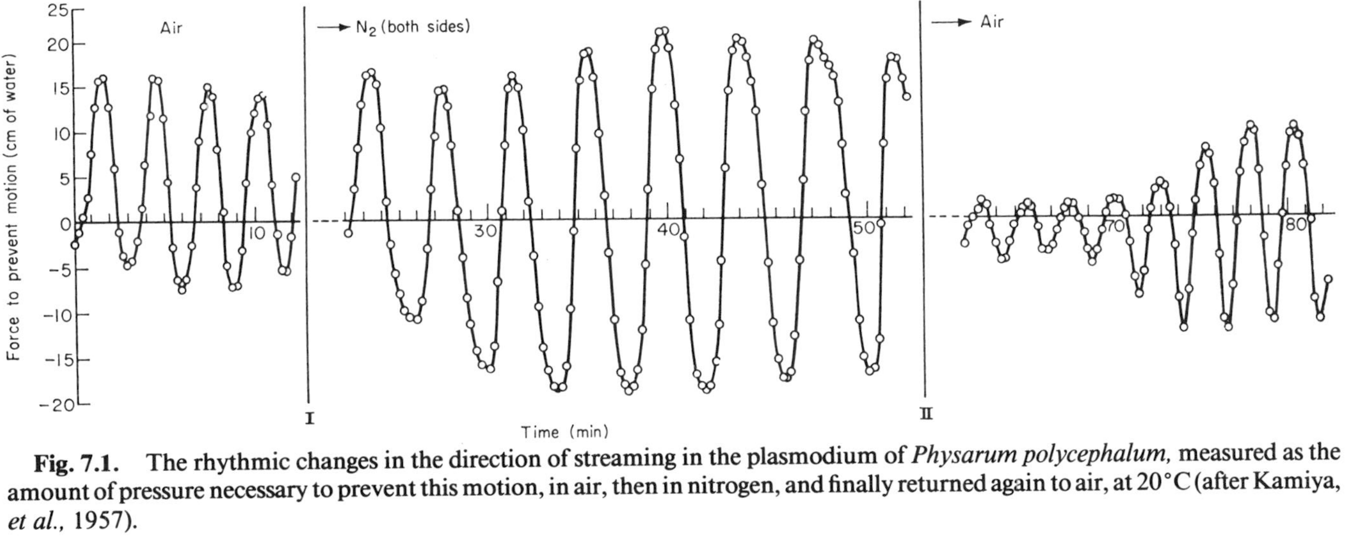

Fig. 7.1. The rhythmic changes in the direction of streaming in the plasmodium of *Physarum polycephalum,* measured as the amount of pressure necessary to prevent this motion, in air, then in nitrogen, and finally returned again to air, at 20°C (after Kamiya, *et al.,* 1957).

ferent effect from KCN, rhythmicity being very much slowed or stopped altogether when this substance was added to both chambers containing the plasmodium. Ether vapor also lengthened the period, and streaming in its presence became erratic. Addition of ATP (2×10^{-3}) to one chamber markedly increased the amplitude of the rhythm but the period was slightly longer in its presence, 3 min as compared with 2 min, before ATP was added at 23°C.

Kamiya (1953) measured the alternation of the direction of plasmodial streaming at a series of different temperatures to obtain information concerning the Q_{10} for both amplitude and frequency of the rhythm. He found that the amplitude was only slightly affected by changes in temperature, while the Q_{10} for frequency was close to 2; thus, the rhythmic aspect of streaming in *Physarum* shows marked temperature dependence. The form of the oscillation is not a perfect sine wave, but its shape can be approximated by combining several sine waves. Different parts of a plasmodium may show oscillations differing in both phase and period. Nothing is known about the factors responsible for the rhythmic alternation in the direction of streaming.

OSCILLATIONS IN GLYCOLYSIS

Rhythms with short periods can be initiated by chemical as well as by electrical stimuli. Betz and Chance (1965) have shown that when yeast (*Saccharomyces carlsbergensis*) is grown aerobically and then transferred to anaerobic conditions with the addition of sugar, the amount of pyridine nucleotide the cells contain oscillates. Observations of this phenomenon were made with a double-beam spectrophotometer capable of determining very rapid changes in pyridine nucleotide absorption as the molecule was alternately reduced and oxidized. The oscillations that could be observed in this way were sinusoidal, although they sometimes varied considerably from simple sine waves. The period *in vivo* was about 0.6 min and was markedly temperature dependent, one preparation showing a period of 37 sec at 25°C and 15 sec at 35 °C. The oscillations damped rapidly in many preparations. In parallel with the reduction of pyridine nucleotide were decreases in ATP, glucose-6-phosphate, and fructose-6-phosphate and increases and fructose diphosphate, all oscillating together.

Extracts that contained the glycolytic enzymes also oscillated *in vitro* but with considerably longer periods than *in vivo* (Chance *et al.*, 1964a,b), again markedly dependent on the temperature, with periods of 7–9 min at 25 °C, 41 min at 12 °C, and 86–98 min at 0 °C. About eight cycles could thus be seen clearly at 0 °C and oscillations continued for 12 h. When oscillation in an extract had disappeared, it could be reinitiated by the addition of the substrate, trehalose (Fig. 7.2). A phase advance of 3–5 min could be induced by the addition of ADP just after the pyridine nucleotide was maximally reduced. This treatment did not change the period. The addition of pyruvate also caused a phase change, this time a delay. Winfree (1972) has studied this system and shown how similar it is in resetting behavior to the circadian rhythm in eclosion in *Drosophila*. The primary site of the generation of the oscillation was deduced to be the regulation of the activity of the enzyme phosphofructokinase by its product, fructose diphosphate, a feedback loop mechanism (Pye, 1973). This enzyme is the rate-limiting step in the flow of glucose through the whole glycolytic pathway.

The oscillations observed in glycolysis in yeast and extracts prepared from it take place in solution. Similar oscillations have also been produced in mitochondria (Chance and Yoshioka, 1966). These oscillations involve the transport of H^+ and K^+ through membranes, and they are generated in the presence of oxygen and the ionophore valinomycin which permits K^+ to move down its concentration gradient through the mitochondrial membranes. A corresponding outflow of H^+ also takes place. Oscillations of these two parameters are 180 ° out of phase, or nearly so, and continue as long as oxygen has not been exhausted by respiration. Oxidative processes in the mitochondrion must be intact for the oscillations to occur and ATP may be a prerequisite, since oligomycin, an inhibitor of the mito-

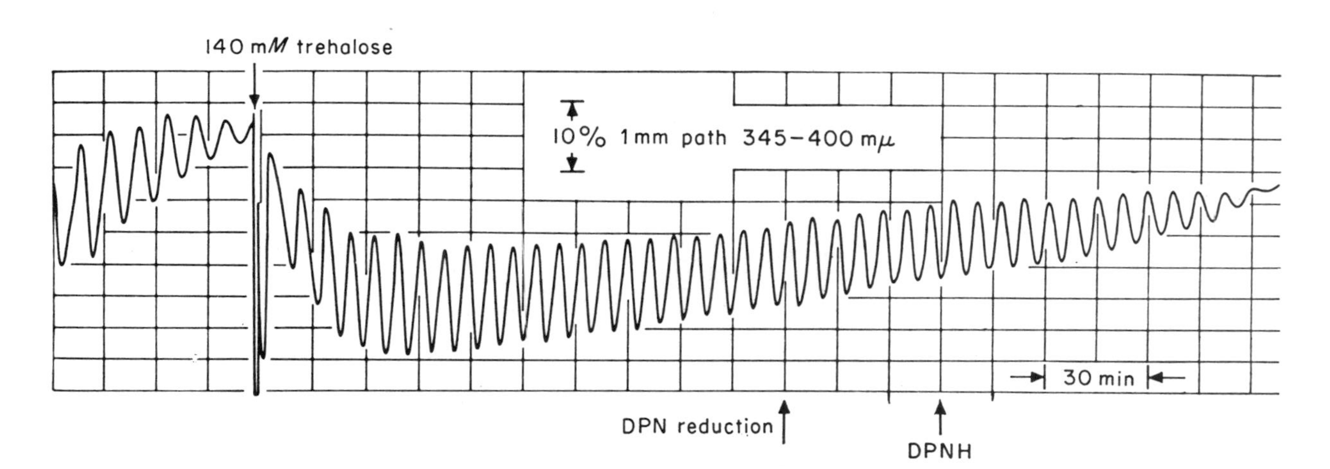

Fig. 7.2. Oscillations in the reduction level of pyridine nucleotide in yeast extracts, sustained by the addition (at arrow) of 140 m*M* trehalose. The period is approximately 7 min at 25°C. Reproduced with permission from Pye and Chance (1966).

chondrial ATPase, stops the rhythm. The exact mechanism by which these oscillations in the mitochondria are generated is not yet known in detail.

OSCILLATIONS WITH SHORT PERIODS IN LEAVES AND ROOTS

Rhythms with periods of the order of minutes are known in leaves of higher plants. These short-period oscillations sometimes appear superimposed on records of circadian leaf movements in plants such as *Phaseolus* (Alford and Tibbitts, 1971). They seem to be especially common during the day phase of circadian rhythms in beans. Darwin measured regular oscillations with a 4- to 5-min period in the leaves of *Averrhoa bilimbi* that were closing in the evening (Darwin and Darwin, 1881). Particularly striking are the movements of the leaves of *Desmodium girans*, which account for the specific name of this plant. The up-and-down motions of the leaflets of *D. girans* have been recorded and appear very regular for at least 4 h, oscillating with a period of about 3 min (Bose, 1927). Guhathakurta and Dutt (1961) showed that electrical oscillations accompanied the changes in leaf position. The movement of leaves are brought about by osmotic changes, then turgor changes in the pulvini in a way similar to circadian sleep movements. Interesting for comparison is the rhythmicity in the degree of opening of stomata reported by Stålfelt (1965). The period of this rhythm is longer, 15–20 min (Hopmans, 1971), and it depends on osmotic swelling of the guard cells. It has been shown that the turgor of the guard cells requires the presence of K^+ and can be correlated with changes in the electrical potential, implicating a potassium pump, the activity of which may be the site of the generation of the oscillation (Fischer, 1968). Short-period rhythms in the membrane potential of 10 sec to 12 min have also been detected in *Ipomoea, Pisum,* and *Xanthium* (Pickard, 1972) and in *Phaseolus* (Aimi and Shibasaki, 1975).

Another oscillation with a period of a few minutes has been found in roots, that of an electrical potential between the tip and the base in dilute salt solution. In the bean root studied by Scott (1957, 1962), Jenkinson and Scott (1961), and Jenkinson (1962a,b), small, apparently spontaneous oscillations with periods of about 5 min can sometimes be found in this potential. These oscillations may persist for hours. Similar rhythms can be induced by stimulating the root electrically, mechanically, or even chemically, but these rhythms damp out very quickly as a general rule. When the root was alternately placed in water and then 33 m*M* sucrose, the electrical potential changed in response to the new osmotic pressure of the medium, but the changes were much greater when the medium was changed at about 5-min intervals than at longer or shorter times, suggesting that 5 min represented the natural resonant frequency of the potential. Roots respond to repetitive changes in the auxin concentration in the medium in a similar manner.

OSCILLATIONS IN THE GROWTH OF PLANTS

Rapidly growing parts of plants, for example the shoots and root tips of seedlings and young tendrils, do not usually grow in a constant direction. When careful records are made of their progress, they are seen to wave back and forth, describing with their tips circles or ellipses of different sizes. Baillaud (1953) has observed the stems of *Cuscuta* to execute this pattern with a period of 5–10 min at 23°C. The movements are very temperature dependent, the period showing a Q_{10} of 2. Galston and coworkers (1964) used a time-lapse camera to observe the growth of the plumules of pea seedlings. When completely etiolated, plants showed little bending, but in continuous red light the plumules rotated with a period of 77 min (Fig. 7.3).

Spurny (1968) reported spiral oscillations in the growing radicle of pea. Hypocotyls of safflower (*Carthamus tinctorius*) also were observed to oscillate but only as a result of geotropic or phototropic stimulation. The period was long compared with other growth rhythms and not very dependent on temperature, being 240 min at 20°C and 210 min at 30°C (Karvé and Salanki,

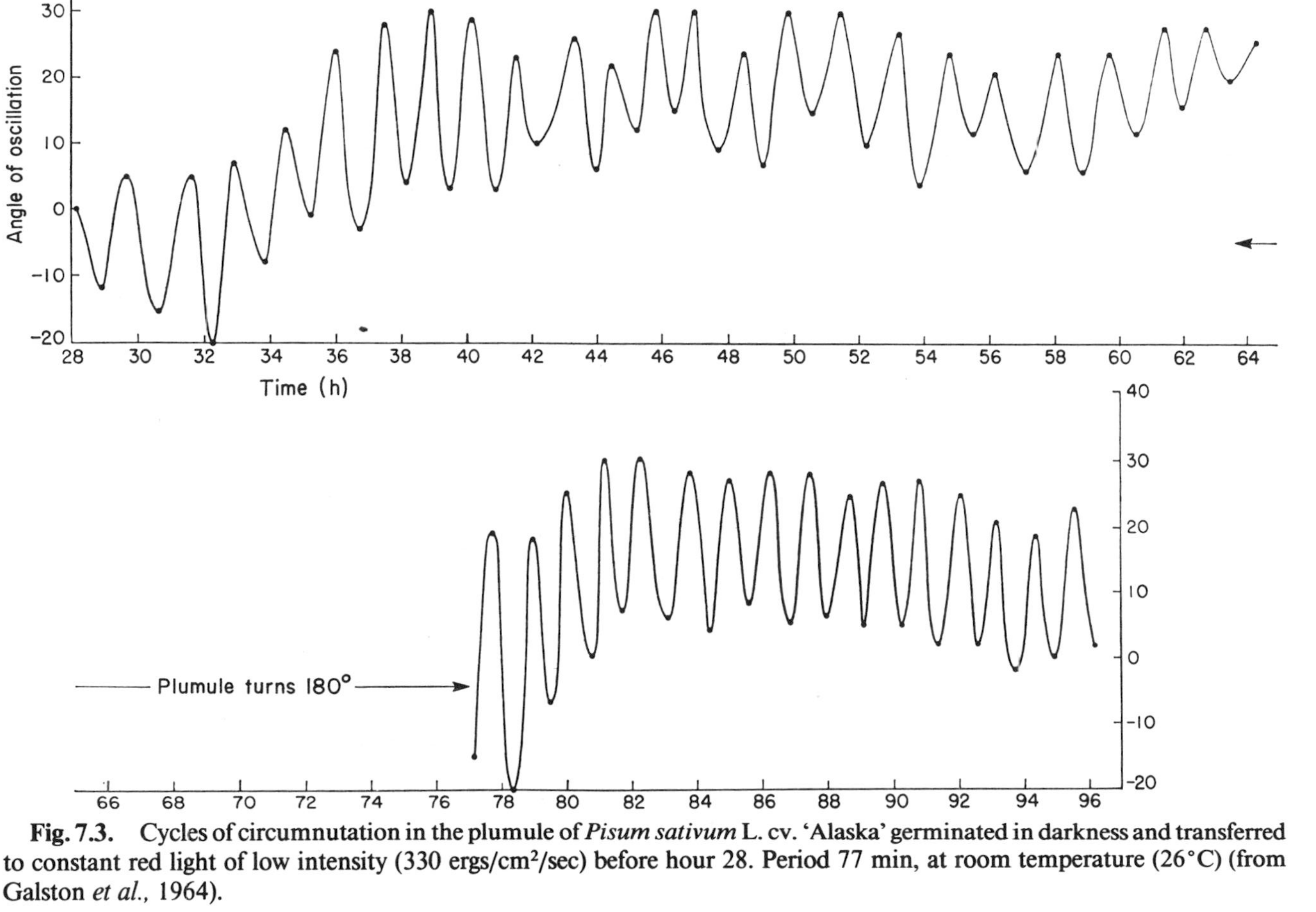

Fig. 7.3. Cycles of circumnutation in the plumule of *Pisum sativum* L. cv. 'Alaska' germinated in darkness and transferred to constant red light of low intensity (330 ergs/cm^2/sec) before hour 28. Period 77 min, at room temperature (26°C) (from Galston *et al.*, 1964).

1964). A second geotropic stimulus initiated a new phase, especially when the stimulus was given at or just after a maximum in the oscillation.

An interesting rhythmicity in the response of seedlings to light has been studied by Klein and his students. When seedlings are germinated in darkness, stems and stem-like structures grow long but leaves remain very small. Irradiation with red light, which converts phytochrome to the far-red-absorbing form, causes a change in development; stem growth slows while leaves grow larger. Gregory and Klein studied the effect of very short flashes of red light on development of previously dark-grown bean seedlings. They found to their surprise that if they irradiated the plants with only two flashes each 2–10 sec long, the increase in weight of the leaves after 24 h more in darkness depended on the time between the two flashes. An oscillation in the effectiveness of the second flash could be detected with a period of 45 sec (Gregory and Klein, 1977, 1978). The marker event for measuring the period was an abrupt increase in leaf weight in 5 sec. The authors postulated that the oscillation was initiated by the first red flash and was a rhythmic change in the response of phytochrome to the second flash. The oscillation was temperature independent between 15 and 30°C. If the first flash was followed by a second irradiation with far-red light, which converts phytochrome back to the inactive red-absorbing form, no oscillations were seen. The weight of radish colyledons similarly treated also varied rhythmically, with a period of 35 sec. Mung bean seedlings apparently behave in a similar way when irradiated with two short pulses of red light separated by different times, although the period is about twice as long as in radish. There is a similar rhythm in the activity of hexokinase and ATPase in mung beans that may be related to the rhythm in sensitivity to light manifested in growth; at least it has the same period of about 60 sec (Mandoli *et al.,* 1978).

Another oscillation with a short period, about 0.5 h, has been discovered in the yeast *Candida utilis* and the amoeba *Acanthamoeba castellanii* (Lloyd *et al.,* 1981, 1982). When the cell division of cultures of these organisms is synchronized, a series of maxima and minima in total protein content and in respiration rate can be seen clearly during the time between divisions. What's more, the period of these oscillations is temperature independent. Only two of the rhythms with short periods have this property of temperature compensation. A number of chronobiologists have suggested that circadian rhythms may represent the interaction of a number of rhythms with much shorter periods than 24 h (see Chapter 4). If this is so, the short oscillations involved would surely be temperature independent.

RHYTHMS WITH PERIODS LONGER THAN A DAY

Fungi show rhythms in growth that under some circumstances appear to have periods of 24 h, but which have properties that set them apart from typical circadian rhythms. Two growth forms alternate when the spores are inoculated at one end of a long tube partially filled with solid medium, and grown in the dark. Unlike circadian rhythms, the period of some of these fungal growth rhythms is temperature dependent and can be as long as 90 h. Grown in the dark on a medium containing yeast extract and a number of amino acids, *Aspergillus ochraceus* and *Colletotrichum lindemuthianum* display a 4-day periodicity in growth pattern at 23°C (Jerebzoff, 1965). Minor constituents of the medium, particularly amino acids, seem to make the difference between zonation and none. These fungal rhythms are not to be confused with the truly circadian rhythm in *Neurospora crassa* band strain, described in Chapter 3. The existence among fungi of rhythms similar in many respects to circadian rhythms yet capable of much longer periods than 24 h, temperature dependent, and not reset by light, is most interesting because they might be considered to represent the ancestral type from which the more sophisticated circadian organization has evolved. This view is strengthened by the presence in fungi of typical circadian rhythms, most thoroughly studied in *N. crassa.*

There is some evidence for the existence of rhythms with a period of 1 week, called by Halberg "circaseptan." In an abstract (Halberg *et al.,* 1985) it is stated that the bioluminescence of

Gonyaulax polyedra varies in intensity with a period of 7 days under certain conditions, but no data are given supporting this assertion.

RHYTHMS WITH PERIODS LONGER THAN A WEEK

Rhythms with periods of months or years are understandably poorly studied. That they do exist is indicated by at least a few observations made under controlled conditions. For example, Lavarennel-Allary (1966) has grown seedlings of several species of oak in controlled environments in LL at 27°C. These seedlings alternately grow and become dormant, as do tree seedlings in nature, showing four to eight such periods in the course of 7 months. In onion bulbs that were stored at 5°C for 2 years and sampled twice each month during this time, Jaffe and Isenberg (1965) reported a periodicity in the rate of elongation of excised leaves of about 2.5 months. Rhythms with a period of about 1 year have been considered in Chapter 6.

RHYTHMS WITH PERIODS OF MANY YEARS

While experiments under controlled conditions have not been attempted, observations in the field indicate that the large tropical bamboos in India and Jamaica flower at intervals of 30–40 years. In parts of Asia a man is said to be so old that he has seen the bamboo flower twice! That this is not the result of environmental cues is indicated by the simultaneous flowering of bamboo derived from the same stock in very different localities—the gardens at Kew and the tropical rain forests of Jamaica, for example (Young and Haun, 1961). Walker (1956) reported that the *Laminaria* population off the coast of Scotland varied in density over a 10-year period, but this variation showed a strong correlation with sun spot activity and thus weather changes are probably responsible rather than a rhythm. The understanding of factors controlling very long cycles, whether endogenous or exogenous, must await long-term cooperative investigations.

REFERENCES

Aimi, R., and Shibasaki, S. (1975). Diurnal changes in bioelectric potential of *Phaseolus* plant in relation to leaf movement and light condition. *Plant Cell Physiol.* **16**, 1157–1162.

Alford, D. K., and Tibbitts, T. W. (1971). Endogenous short-period rhythms in the movements of unifoliate leaves of *Phaseolus angularis* Wright. *Plant Physiol.* **47**, 68–70.

Baillaud, L. (1953). Action de la temperature sur la periode de nutation des tiges volubiles de Cuscute. *C. R. Acad. Sci. Paris* **236**, 1986–1988.

Betz, A., and Chance, B. (1965). Phase relationship of glycolytic intermediates in yeast cells with oscillatory metabolic control. *Arch. Biochem. Biophys.* **109**, 585–594.

Bose, J. C. (1927). "Plant Autographs and Their Revelations." Longmans, Green, London.

Chance, B., and Yoshioka, T. (1966). Sustained oscillations of ionic constituents of mitochondria. *Arch. Biochem. Biophys.* **117**, 451–465.

Chance, B., Esterbrook, R. W., and Ghosh, J. (1964a). Damped sinusoidal oscillations of cytoplasmic reduced pyridine nucleotide level in yeast cells. *Proc. Natl. Acad. Sci. U. S. A.* **51**, 1244–1251.

Chance, B., Schoener, B., and Elsaesser, S. (1964b). Control of the waveform of oscillations of the reduced pyridine nucleotide level in a cell-free extract. *Proc. Natl. Acad. Sci. U. S. A.* **52**, 337–341.

Darwin, C., and Darwin, F. (1881). "The power of movement in plants." Appleton, New York.

Fischer, R. A. (1968). Stomatal opening: Role of potassium uptake of guard cells. *Science* **160**, 784–785.

Galston, A. W., Tuttle, A. A., and Penny, P. J. (1964). A kinetic study of growth movements and photomorphogenesis in etiolated pea seedlings. *Am. J. Bot.* **51**, 853–858.

Gregory, A., and Klein, A. O. (1977). Phytochrome-initiated fast rhythm controlling developmental response. *Nature (London)* **265**, 335–337.

Gregory, A., and Klein, A. O. (1978). An oscillating system regulating development in plants. *Photochem. Photobiol.* **27**, 133–136.

Guhathakurta, A., and Dutt, B. K. (1961). Electrical correlate of the pulsatory movement of *Desmodium girans*. *Transl. Bose Res. Inst.* **24**, 73–82.

Halberg, F., Hastings, W., Cornelissen, G., and Broda, H. (1985). *Gonyaulax polyedra* 'talks' both 'circadian' and 'circaseptan'. *Chronobiologia (Milan)* **12**, 185 (abstr.).

Hopmans, P. A. M. (1971). Rhythms in stomatal opening of bean leaves. *Meded. Landbouwhogesch. Wageningen* **71-3**, 1–86.

Jaffe, M. J., and Isenberg, F. M. R. (1965). Rhythmic growth in excised sprout-leaves of onion bulbs. *Plant Physiol.* **40**, Suppl. xx (abstr.).

Jenkinson, I. S. (1962a). Bioelectric oscillations in bean roots: Further evidence for a feedback oscillator. II. *Aust. J. Biol. Sci.* **15**, 101–114.

Jenkinson, I. S. (1962b). Bioelectric oscillations in bean roots: Further evidence for a feedback oscillator. III. *Aust. J. Biol. Sci.* **15**, 115–125.

Jenkinson, I. S., and Scott, B. I. H. (1961). Bioelectric oscillations of bean roots: Further evidence for a feedback oscillator. I. *Aust. J. Biol. Sci.* **14**, 231–247.

Jerebzoff, S. (1965). Manipulation of some oscillating systems in fungi by chemicals. *In* "Circadian Clocks" (J. Aschoff, ed.), pp. 183–189. North-Holland Pub. Amsterdam.

Kamiya, N. (1953). The motive force responsible for protoplasmic streaming in the myxomucete plasmodium. *Annu. Rep. Fac. Sci., Osaka Univ.* **1**, 53–83.

Kamiya, N. Nakajima, H., and Abe, S. (1957). Physiology of the motive force of protoplasmic streaming. *Protoplasma* **48**, 94–112.

Karvé, A. D., and Salanki, A. S. (1964). Effect of alcohol and light upon geotropically induced oscillations. *Zeitsch. f. Bot.* **52**, 186–192.

Lavarenne-Allary, S. (1966). Croissance rhythmique des quelques espèces de Chêne cultivées en chambres climitisées. *C. R. Hebd. Seances Acad. Sci., Ser. D* **262**, 358–361.

Lloyd, D., Edwards, S. W., and Williams, J. L. (1981). Oscillatory accumulation of total cellular protein in synchronous cultures of *Candida utilis. FEMS Microbiol. Lett.* **12**, 295–298.

Lloyd, D., Edwards, S. W., and Fry, J. C. (1982). Temperature-compensated oscillations in respiration and cellular protein content in synchronous cultures of *Acanthamoeba castellanii. Proc. Natl. Acad. Sci. U. S. A.* **79**, 3785–3788.

Mandoli, D. F., José, A. M., Fischer, C., and Klein, A. O. (1978). Rapid oscillations modulating enzyme activity and growth in etiolated seedlings. *Proc. Annu. Eur. Symp. Photomorphog.*, pp. 67–68.

Pickard, B. G. (1972). Spontaneous electrical activity in shoots of *Ipomoea, Pisum* and *Xanthium. Planta* **102**, 91–114.

Pye, E. K. (1973). Glycolytic oscillations in cells and extracts of yeast, some unsolved problems. *In* "Biological and Biochemical Oscillators" (B. Chance, A. K. Ghosh, E. K. Pye, and B. Hess, eds.), pp. 269–284. Academic Press, New York.

Pye, E. K., and Chance, B. (1966). Sustained sinusoidal oscillations of reduced pyridine nucleotide in cell-free extracts of *Saccharomyces carlbergensis. Proc. Natl. Acad. Sci. U. S. A.* **55**, 888–894.

Scott, B. I. H. (1957). Electric oscillations generated by plant roots and a possible feedback mechanism responsible for them. *Aust. J. Biol. Sci.* **10**, 164–170.

Scott, B. I. H. (1962). Feedback-induced oscillations of five-minute period in the electric field of the bean root. *Ann. N. Y. Acad. Sci.* **98**, 890–900.

Spurny, M. (1968). Spiral oscillations of the growing radicle in *Pisum sativum* L. *Naturwissenschaften* **55**, 46.

Stålfelt, M. G. (1965). The relation between the endogenous and induced elements of the stomatal movements. *Physiol. Plant.* **18**, 177–184.

Walker, F. T. (1956). Periodicity of the Laminariaceae around Scotland. *Nature (London)* **177**, 1246.

Winfree, A. T. (1972). Oscillatory glycolysis in Yeast: The pattern of phase resetting by oxygen. *Arch. Biochem. Biophys.* **149**, 388–401.

Yates, G. T. (1986). How microorganisms move through water. *Am. Sci.* **74**, 358–365.

Young, R. A., and Haun, J. R. (1961). Bamboo in the United States: Description, culture, utilization. *U.S. Dep. Agric., Agric. Handb.* **193**.

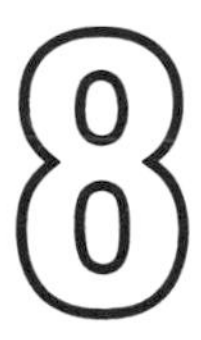

Biological Clocks and Human Affairs

Flowers have clocks, bees have clocks; do we humans have a biological clock? The answer is "yes." Although we do not use it very often, we have a perfectly good circadian clock. Our time sense has been corrupted by our use of mechanical watches and other time pieces and by our artificial lights, but sometimes we awake just before the alarm sounds and wonder how we did it.

As in other organisms, the presence or absence of a circadian rhythmicity in humans can only be demonstrated by measurements in a constant environment without time cues. Such experiments require very special equipment, including living quarters where there is no chance of the subject obtaining information about the time from any external source. Automatic recording of physiological parameters without the intervention of the subject are also necessary. For studies of the circadian rhythms of humans, such living quarters were constructed by Aschoff and Wever at the Max Planck Institüt für Verhaltenphysiologie, in Andechs near Munich; they were built underground so that even the sound of the church bells could not be heard. Here body temperature, sleeping and activity, and urine volume and chemistry were recorded automatically for 147 volunteers. The results were clear: all subjects showed rhythms in these properties. The periods were not 24 h but usually about 25 h, in a few cases as long as 27 h (Wever, 1979). A record from the work of Wever showing the sleep and body temperature rhythms of one subject is reproduced as Fig. 8.1. In subects living alone in one of these isolation units under constant conditions of LL of two different light levels or in DD, the periods were very much the same (Aschoff and Wever, 1981). The fact that the period of a given volunteer did not vary, even when the light intensity in LL was changed or darkness was substituted, and was always longer than 24 h is strong evidence for the existence of a circadian clock in humans, very similar to that found in other mammals and even in flagellates like *Gonyaulax.*

Many physiological and behavioral properties have been found to vary over the day and night in humans. Alertness and speed in doing calculations is greatest during the day. Hormone levels change diurnally as does our sensitivity to drugs. There are more births and deaths at night than during the day. However, we do not know whether or not all of these changes are true circadian rhythms and not direct effects of night and day, since experiments in many cases have not and cannot be done on humans in a constant environment suite like that at Andechs. Observations of animals, including monkeys, under constant conditions have strongly suggested that these processes are manifestations of circadian timing in humans.

While normally we humans are unaware that we possess a circadian clock, there are occasions when our biological timekeeping becomes evident because there is a conflict between our body time and environmental time. One such circumstance is during east–west air travel. Airplanes fly so fast that passengers are transported across a number of time zones within a few hours. If I were to fly from California to Paris today, I would cross eight time zones and arrive to find the time there 9 h later than my body time. Eventually the sequence of day and night in Paris and other time cues would reset my biological clock, but not all the physiological parameters controlled by this clock would be shifted equally rapidly. There might follow a time

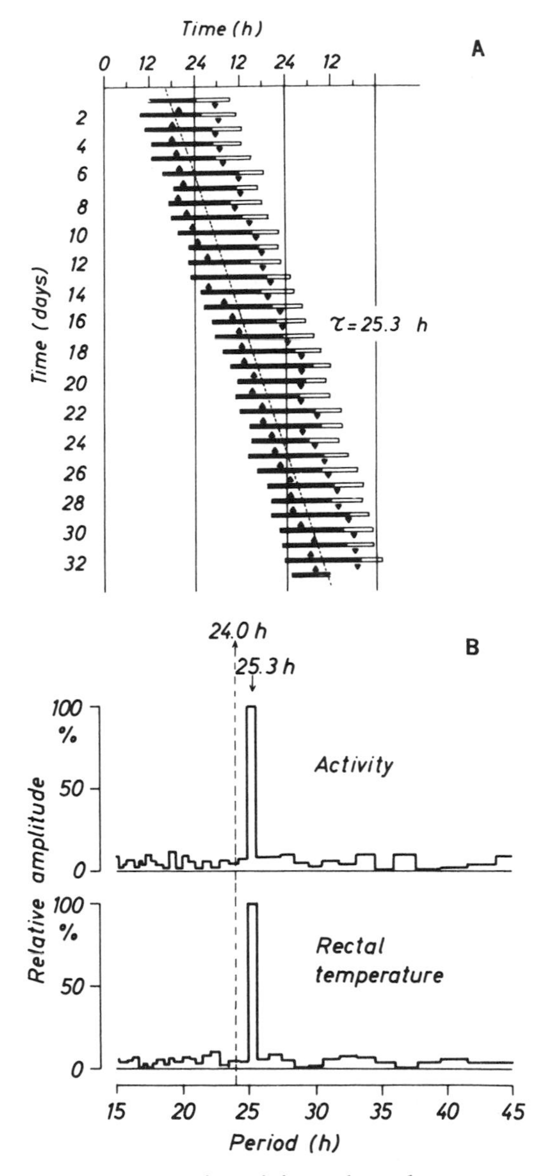

Fig. 8.1. Circadian rhythms in activity and rectal temperature of a subject living under constant conditions at Andechs. A. Successive periods plotted one under the other; black bars indicate activity and white bars indicate rest; ▲ = maximum body temperature, ▼ = minimum body temperature. B. The same data plotted as a periodogram, showing a marked peak at 25.3-h period in both activity and body temperature (Wever, 1979, Fig. 16).

when I wanted to sleep during the day and was incapable of thinking clearly or making decisions during the day, perhaps also suffered from headaches and insomnia at night. This condition is known as "jet-lag." These symptoms are not caused by fatigue during air travel, because they do not occur during north–south flights where the time remains the same at the destination as at the point of departure.

What can we do to lessen or prevent jet-lag? So far there is unfortunately no "reset pill" that we can take. However, knowledge of the resetting of circadian rhythms suggests that we try either to avoid resetting or to hasten it. Avoiding a shift in biological timing by following the home schedule in the new location can only be useful for quick visits abroad, a day or two at most. However it may be possible to shorten the time required for resetting. It has been found that jet-lag lasts a shorter time when the traveler is out of doors and active than when he stays isolated in his hotel room in artificial light (Moore-Ede *et al.,* 1982). Strong resetting stimuli, such as a definite light–dark cycle and time-dependent social cues, will hasten resetting of biological timing. There is now evidence that the circadian rhythms in humans, like those in other organisms, are reset by light (Czeisler *et al.,* 1986). Bright light, more than 2500 lux, appears to be required. There are a number of drugs that are capable of resetting circadian rhythms in plants and animals, as we have seen. Most of these are so toxic that they would be dangerous for humans to use to reset their clock when starting a long east–west trip. However, very recently benzodiazepine, a drug used in the treatment of human insomnia, was found to reset the circadian activity rhythm of another mammal, the golden hamster (Turek and Losee-Olson, 1986). There is hope that a pill containing this substance may soon be available for jet-lag.

Because it is more efficient to operate some industrial plants and emergency services continuously, it has become common to employ workers at night as well as during the day. People who work the night shift over long periods of time become adjusted to their reversed schedule and are able to sleep during daytime. However, they may be isolated from the social life of their families and friends or try to adapt to a normal schedule every weekend. For this reason, it is sometimes the practice in industry to change workers from night shift to day shift and back at intervals. Circadian principles have not been considered in planning how often to change the work schedule from night to day and back. A study was made of the sleep and general well-being of workers at a company where some were working in shifts that changed every 7 days while others worked only during the day shift (Czeisler *et al.,* 1982). As compared with the latter, night-shift workers complained of more insomnia and reported that they had fallen asleep at work at least once during the last 3 months. They reported that they required 2–4 days or longer to become accustomed to a new shift, as would be expected for resetting their circadian sleep–wake cycle and other rhythms. What we know about circadian time-keeping suggests that changing shifts as often as once a week is too often. In the industry under study, a 3-week interval between changing shifts was tried, on the theory that this would allow time for resetting the workers' rhythms. It is thought that humans with their natural period longer than 24 h can shift their timing more easily by a delay than by an advance, i.e., to a later rather than to an earlier time. Taking these two considerations into account resulted in greater efficiency and contentment.

Not considering the timing of the sleep–wake cycle can have dire effects that should concern us all, although we may not ourselves work at night. For example, in assigning flight schedules to pilots, their sleep cycle may not be taken into account so that they fly for several days in a row during the time when their clock tells them to sleep. One pilot of a 727 airliner that was coming in for a landing at the Los Angeles International Airport fell asleep when the plane was at only 200 feet above ground; a crash was only avoided when the co-pilot noticed what was happening and flew the plane to a safe landing (Moore-Ede *et al.,* 1982, p. 333). In another plane coming into Los Angeles after a long flight, the entire flight crew fell asleep and could not be roused until the plane had overshot Los Angeles and was 100 miles out over the Pacific. Another plane was not so fortunate. It crashed in Bali killing 96 passengers and 11 crew members after the personnel had flown during several sleep cycles. People responsible for monitoring atomic energy plants during the night may also fail to reset their sleep–wake cycle and so be prone to error at 2–4 A.M., the time, by the way, when trouble occurred at Three Mile Island and Chernobyl.

It is known now that the sensitivity of animals, insects, and humans to a number of drugs changes with time of day with the characteristics of a circadian rhythm. This has some interesting practical applications. For example, since some insect pests are more sensitive to insecticides at certain times of day than at others, it is possible to use a smaller amount if the poison is applied at the right time and hence avoid a part of the environmental pollution with these chemicals. On the other hand, giving medicines to patients at the most sensitive time allows the dose to be smaller and reduces unpleasant side effects. This principle has already been applied with good results in the chemotherapy of human cancers.

If you are interested in learning more about the circadian rhythms in humans and the application of our knowledge of these phenomena to medicine, you will find additional information in the references that follow (Minors and Waterhouse, 1981; Palmer, 1976; Reinberg and Smolensky, 1983; Reinberg *et al.*, 1984).

At the present time, not much is known about rhythms in humans with frequencies that do not match the day–night cycle. Monthly cycles obviously are important in women, but these menstrual cycles are not synchronized with any natural periodicity, the phase of the moon, or the related tidal variations, for example. There are hints of yearly cycles of birth and death in humans (Aschoff and Wever, 1981), but these cycles are very difficult if not impossible to study in an environment without external yearly variations. Much more research will have to be done before such cycles are understood.

In summary, the study of rhythmic phenomena in plants has stimulated similar work with animals. The resemblance between time-keeping in otherwise very different organisms has proven to be striking. Thus, insights from studies of plant rhythms can be applied to animals and to ourselves, not to mention our crop plants on which we ultimately depend for our food.

REFERENCES

Aschoff, J., and Wever, R. (1981). The circadian system of man. *In* "Biological Rhythms" (J. Aschoff, ed.), pp. 311–331. Plenum, New York.

Czeisler, C. A., Moore-Ede, M. C., and Coleman, R. M. (1982). Rotating shift work schedules that disrupt sleep are improved by applying circadian principles. *Science* **217**, 460–463.

Czeisler, C. A., Allan, J. S., Strogatz, S. H., Ronda, J. M., Sanchez, R., Rios, C. D., Freitag, W. O., Richardson, G. S., and Kronauer, R. E. (1986). Bright light resets the human circadian pacemaker independent of the timing of the sleep-wake cycle. *Science* **233**, 667–671.

Minors, D. S., and Waterhouse, J. M. (1981). "Circadian Rhythms and the Human." Wright, Bristol, England.

Moore-Ede, M. C., Sulzman, F. M., and Fuller, C. A. (1982). "The Clocks that Time Us." Harvard Univ. Press, Cambridge, Massachusetts.

Palmer, J. D. (1976). "An Introduction to Circadian Rhythms." Academic Press, New York.

Reinberg, A., and Smolensky, M. H. (1983). "Biological Rhythms and Medicine." Springer-Verlag, Berlin and New York.

Reinberg, A., Smolensky, M., and Labrecque, G., eds. (1984). Biological rhythms and medication. *Annu. Rev. Chronopharmacol.* **1**,1–422.

Turek, F. W., and Losee-Olson, S. (1986). A benzodiazepine used in the treatment of insomnia phase-shifts the mammalian circadian clock. *Nature (London)* **321**, 167–168.

Wever, R. A. (1979). "The Circadian System of Man. Results of Experiments under Temporal Isolation." Springer-Verlag, Berlin and New York.

Author Index

Subject Index